Farm Structures and Equipment

With Information on the Farmhouse, Wells, Water Piping, Heating Systems and Livestock Houses

By

W. C. Krueger

British Library Cataloguing-in-Publication Data
A catalogue record for this book is available from the
British Library

Agricultural Tools and Machinery

Farming has an incredibly long history. Beginning around 3000 BC, nomadic pastoralism, with societies focused on the care of livestock for subsistence, appeared independently in several areas in Europe and Asia. This form of farming utilised basic implements, but with the rise of arable farming, agricultural tools became more intricate. Between 2500 and 2000 BC, the simplest form of the plough, called the ard, spread throughout Europe, replacing the hoe (simply meaning a 'digging stick'). Whilst this may not seem like a revolutionary change in itself, the implications of such developments were incredibly far reaching. This change in equipment significantly increased cultivation ability, and affected the demand for land, as well as ideas about property, inheritance and family rights.

Tools such as hoes were light and transportable; a substantial benefit for nomadic societies who moved on once the soil's nutrients were depleted. However, as the continuous cultivating of smaller pieces of land became a sustaining practice throughout the world, ploughs were much more efficient than digging sticks. As humanity became more stationary, empires such as the New Kingdom of Egypt and the Ancient Romans arose, dependent upon agriculture to feed their growing populations. As a result of intensified agricultural practice, implements continued to improve, allowing the expansion of available crop varieties, including a wide range of fruits, vegetables, oil crops, spices and other products. China was also an important centre for agricultural technology development during this period. During the Zhou dynasty (1666–221 BC), the first canals were built, and irrigation was used extensively. The later Three Kingdoms and Northern and Southern dynasties (221–581 AD) brought the first biological pest control, extensive writings on agricultural topics and technological innovations such as steel and the wheelbarrow.

By 900 AD in Europe, developments in iron smelting allowed for increased production, leading to improved ploughs, hand tools and horse shoes. The plough was significantly enhanced, developing into the mouldboard plough, capable of turning over the heavy, wet soils of northern Europe. This led to the clearing of forests in that area and a significant increase in agricultural production, which in turn led to an increase in population. At the same time, farmers in Europe moved from a two field crop rotation to a three field crop rotation in which one field of three was left fallow every year. This resulted in increased productivity and nutrition, as the change in rotations led to different crops being planted, including vegetables such as peas, lentils and beans. Inventions such as improved horse harnesses and the whippletree (a mechanism to distribute force evenly through linkages) also changed methods of cultivation. The prime modes of power were animals; horses or oxen, and the elements; watermills had been initially developed by the Romans, but were significantly improved throughout the Middle Ages, alongside windmills – used to grind grains into flour, cut wood and process flax and wool, among other uses.

With the coming of the Industrial Revolution and the development of more complicated machines, farming methods took a great leap forward. Instead of harvesting grain by hand with a sharp blade, wheeled machines cut a continuous swath. And instead of threshing the grain by beating it with sticks, threshing machines separated the seeds from the heads and stalks. Perhaps one of the most important developments of this era was the appearance of the tractor; first used in the late nineteenth century. Power for agricultural machinery could now come from steam, as opposed to animals, and with the invention of steam power came the portable engine, and later the traction engine; a multipurpose, mobile energy source that was the ground-crawling cousin to the steam locomotive. Agricultural steam engines took over the heavy pulling work of horses, and were also equipped with a pulley that could power stationary machines via the use of a long belt. They did operate at an incredibly slow speed however, leading farmers to amusingly comment that tractors had two speeds: 'slow, and damn slow.'

From this point onwards, it has been the methods of powering machines, rather than the agricultural machines themselves, which have been the biggest breakthroughs in farming practice. The internal combustion engine; first the petrol engine and later diesel engines, became the main source of power for the next generation of tractors. These engines also contributed to the development of the self-propelled, combined harvester and thresher, or 'combine harvester' (also shortened to 'combine'). Instead of cutting the grain stalks and transporting them to a stationary threshing machine, these combines cut, threshed, and separated the grain while moving continuously through the field. Combines might have taken the harvesting job away from tractors, but tractors still do the majority of work on a modern farm. They are used to pull various implements – machines that till the ground, plant seed and perform other tasks. Besides the tractor, other vehicles have been adapted for use in farming, including trucks, airplanes and helicopters, for example to transport crops and equipment, aerial spraying and livestock herd management.

The basic technology of agricultural machines has changed little in the last century. Though modern harvesters and planters may do a better job or be slightly tweaked from their predecessors, todays combine harvests still cut, thresh, and separate grain in essentially the same way. However, technology is changing the way that humans operate the machines, as computer monitoring systems, GPS locators, and self-steer programs allow the most advanced tractors and implements to be more precise and less wasteful in the use of fuel, seed, or fertilizer. In the foreseeable future, there may be mass production of driverless tractors, and new advances in nanotechnology and genetic engineering are being used in the same way as machines, to perform agricultural tasks in unusual new ways. Agriculture may be one of the oldest professions, but the development and use of machinery has made the job title of *farmer* a rarity. Instead of every person having to work to provide food for themselves, in America for example, less than two percent of the population works in agriculture. But today, a single farmer can produce cereal to feed over one thousand people. With continuing advances in agricultural machinery, the role of the farmer continues on.

FARM STRUCTURES AND EQUIPMENT

W. C. KRUEGER

Professor W. C. Krueger is an "inveterate fixer" who has many opportunities to indulge this proclivity in his position as Extension Service agricultural engineer at Rutgers. In this capacity, and for 15 years, he has advised New Jersey farmers on the building of the farm home and farm structures, and the selection, use, and care of machinery and equipment. Farm buildings now standing in many parts of the state first took definite form on Professor Krueger's drawing board. He has long been active in the promotion of rural electrification and the installation of efficient systems of drainage and irrigation where needed. He is the inventor of an electric soil "pasteurizer." A native of the Badger State, Professor Krueger graduated from the University of Wisconsin, where he was one of the first undergraduates to major in agricultural engineering. He taught that subject at his Alma Mater and, in addition, was in charge of rural electrification research at the University for three years. His record includes four years of teaching at the University of Tennessee. Professor Krueger has served as chairman and vice-chairman of the North Atlantic Section, American Society of Agricultural Engineers.

FARMING is unique in combining the pleasures and advantages of a country home with one's profession. Almost inseparable from other considerations in the choice of a farm is the farmhouse. This is home and office, the center of social and business activity, a visual expression of the owner's tastes, standards, and ideals. Near it are grouped the buildings necessary to farming operations, the entire unit—house, outbuildings, yard, lawn, garden—constituting the farmstead.

Convenient and time-saving arrangement of farm buildings is as important to farm efficiency as factory and machine arrangement to industrial efficiency. Studies based upon a number of otherwise comparable dairy farms show that on the average well-planned dairy farm the operator walks approximately 100 miles a month in doing the daily chores, whereas on poorly arranged farmsteads of equal size the operator walks an average of 135 miles per month. The best farm building arrangement

1

included in these studies gives a record of only 49 miles per month. The location of buildings, their relative positions, and their physical condition and suitability for their respective purposes are therefore important considerations in buying a new farm. It is well to remember that each type of farming will determine its own best type of structures and that the possibility of a shift from one type of agriculture to another must be considered in planning buildings for the future.

While a perfect farmstead layout is rarely achieved, certain fundamental principles apply to all. Among these is the arrangement of buildings in the form of a hollow court giving access to all of the buildings from a main circular driveway. This eliminates crossing fences and the opening of gates and presents an orderly appearance. A compromise is necessary between close grouping of buildings for labor efficiency and dispersal for fire protection. A minimum spacing of fifty feet is recommended. Farmstead drainage is another important consideration. The buildings, and especially the dwelling, should be on high ground, on a porous type of soil. The windows of the kitchen, the work-room of the house, should command a view of the other farm buildings, if possible. On dairy and stock farms, the distance from house to barns should be at least 150 feet, more if prevailing winds blow from barn to house. It is also desirable that the house be at least 100 feet from the highway. Although the most efficient farmstead location is at the center of the farm area, maintenance of private roads and lanes often presents a difficult problem, particularly as to snow removal and the construction of an all-weather roadbed.

The groupings of farm buildings should be given some thought. The hog house, farthest from the dwelling, is best next to the corn crib and the feeding lot. Machine shed, garage, and farm shop constitute another unit. The granary or feed storage can be common to the dairy barn and poultry houses. The lawn, garden, and home orchard, if any, are grouped with the dwelling unit thus facilitating fence exclusion of poultry and livestock.

THE FARMHOUSE

The new owner of a farm home often has difficulty in deciding whether to remodel and recondition the existing farmhouse, or to raze it and rebuild. If the house has historic or antique value, restoration expenditures exceeding the cost of a new house may be justified. If the house has good architectural lines, is reasonably satisfactory from the standpoint of room layout, is in sound structural condition, and is satisfactorily located with respect to the highway and other farm buildings, serious consideration should be given to reconditioning and using it.

Farmhouse as purchased. See opposite page for same house as enlarged by new owner.

Apart from historical or sentimental associations, the first question to ask is, "Is the building worth saving?" The cost of remodeling a run-down house, making it livable, and adding all of the modern conveniences so necessary to enjoyment of a farm home often exceeds the cost of a new structure. Beyond general arrangement and construction, the most revealing points of inspection for structural soundness are the foundation walls and the sills, girders, joists, and beams that make up the main supporting framework for the entire house. Masonry or concrete supports should carry all wood construction members at least 6 inches above ground level. If this is not the case, the remodeling job will involve raising of the structure, replacement of rotted sills and beams in contact with the damp earth, and additions

3

to the foundation walls to bring them up to the required height. All this is likely to cost considerable money for hired help, since the work is beyond the capacity of the average home owner.

Cracked and bulging foundation walls are signs of a poor footing, indicative of uneven settling which may have warped the house framework. Stains and streaks on the inside basement wall are an indication of leaks and a wet basement. "Weeping" walls are hard to treat from the inside surface, and effective outside treatment necessitates exposure down to the footing to permit the application of a waterproof cement or bituminous shield on the.

Remodeled farmhouse. See opposite page for same house before remodeling.

entire wall surface. The mortar joints in stone masonry foundation walls common to old houses may have become crumbly and loose. Inspection with a chisel or a scratch awl will reveal the extent to which joint repointing must be done.

The junction between foundation walls and the house framework is probably the most common point of house failure. The soundness of beams and girders can be determined by probing with an ice pick or knife blade. Look particularly at contact points between wood and masonry walls. Beam ends imbedded in masonry construction are likely to rot for lack of ventilation. Poorly ventilated areas under a house should be given special attention. Enclosed porches and unexcavated portions of the house are likely to show supporting-beam failure.

A probing inspection of all wood construction supporting the

4

house will also reveal the possible presence of termites. These insects, often called white ants, live on wood, always working from within the beam or board to leave a thin outside shell for light exclusion and protection. Since contact with moist earth must be maintained for life, termites can be eradicated or excluded by placing a metal shield between the wood frame or supporting members and the concrete or masonry foundation on which they rest.

Incidentally, the wall sills supporting the house frame should be bolted to the foundation at 8- to 10-foot intervals. These anchor bolts are safeguards against high winds, which have blown many unanchored farm buildings and houses off their foundations, as during the New England hurricane.

Many old houses have outside walls and even partition walls of brick-fill construction or of solid masonry. Such construction makes installation of electric service and plumbing difficult and costly. Such houses are usually hard to heat, expensive to insulate, and difficult to treat to prevent excessive absorption of driving rains by the outside wall surfaces, with resultant interior dampness.

An inspection of the attic will reveal the condition of the roof framing and the extent of necessary repairs. Neglected roof leaks will soon rot rafters and roof boards. Such members are difficult to replace without expensive and extensive operations.

Chimney safety and performance should never be taken for granted. Crumbling mortar joints in unlined chimney flues constitute a fire menace. The junctions of chimney with roof line and attic floor are particularly likely to cause trouble. These points may be inspected from the attic. Leaks in a chimney may be tested by placing a smudge of oily rags or waste in the lowest chimney opening after having stopped up the chimney at the top. Smoke escaping into the rooms or through the sides of the chimney is a warning that a fire hazard exists. While reconstruction of a chimney does not involve unusual difficulties or excessive costs, the necessity of this work should be determined as a factor in purchasing, remodeling, or rebuilding. For fire protection and in compliance with most state building codes, standard flue lining should be used in all chimneys. Defective chimneys are listed as one of the leading causes of farm fires; with farm fires in the United States occurring on an average of every fifteen minutes the year round, the importance of safe chimney construction is emphasized.

Many new owners, after reasonably careful surveys and inspections, have committed themselves to a remodeling program only to find that unforeseen though necessary repairs and other difficulties have multiplied the original cost estimate many times.

Look for masonry failure between chimney bricks; unlined flues are a fire hazard.

Since labor is the chief factor in such programs, owners who are skilled in carpentry, plumbing, and cabinet making, and who are generally handy with tools, can, over a period of years, accomplish a complete schedule of restoration with a reasonable expenditure. Such gifts are rare, however, and the average home owner will do well to rely on a competent architect for planning and on a contractor for actual construction. It is well known that an architect can save more than his fee in reduced construction and materials' costs and that the lowest priced worker or contractor often proves the most expensive.

WATER SUPPLY

No farmstead utility exceeds in importance a dependable, safe, and palatable water supply. The pioneer and homesteader, when-

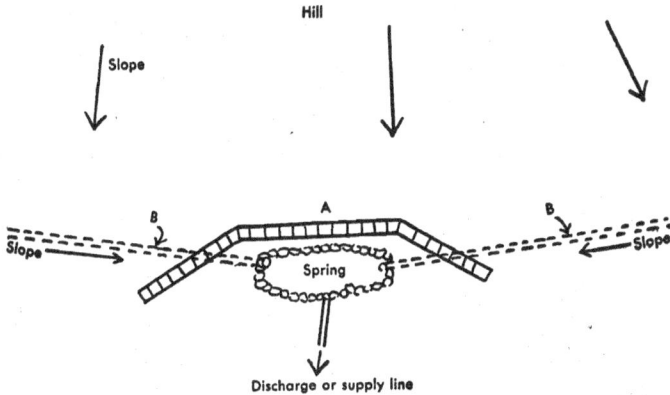

SPRING WATER SUPPLY

A—*Concrete, brick, or masonry wall above spring to divert surface water and trash.*
B—*Field or agricultural tile laid at ground-water level to intercept water that otherwise would by-pass the spring. Tile discharges into spring basin or sump, augmenting normal rate of flow. This arrangement should be used only when area above spring is free from contamination.*

ever possible, located his home site near a spring or in an area where ground water reservoirs were within easy reach. Then, as now, water was all-important.

Springs

Conversion of forests to farm lands, clean tillage practices, and the construction of vast drainage projects have caused many springs to fail and soil water tables to lower. Those springs that remain are often endangered by surface contamination, particularly in our highly populated areas. No spring, no matter how clear and sparkling it may appear, can be considered safe unless periodic bacteriological tests are made to determine purity. Under most soil conditions, ownership control of the area feeding the spring makes it possible to guard against contamination. In limestone rock areas, however, underground streams may travel for

7

miles before emerging as a spring. In such cases, the source of contamination is difficult, if not impossible, to discover or control.

Springs, even though unfit as a potable water supply, can

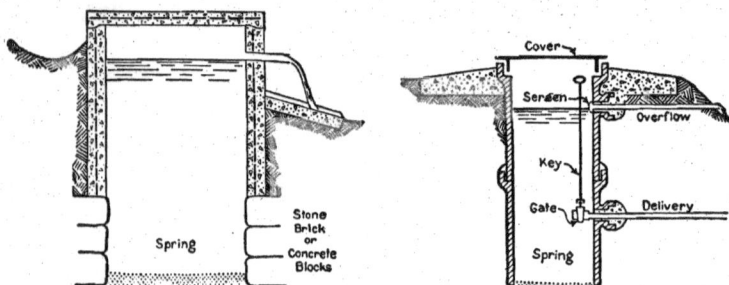

Protection of springs. Curbed and covered to keep out surface wash and to prevent dipping or bailing; water should be drawn only by natural flow through a pipe or by pumping; on the left, square concrete box having 4 or 5 inch walls and 3 inch top reinforced with heavy wire netting or stock fencing; on the right, curb composed of large-size clay or concrete pipe (T branches). (From U.S.D.A. Farmers' Bul. 1448.)

often be made an attractive and useful feature of the farmstead. A sightly spring house of concrete or masonry construction can be the center or head of a landscape area combining a small pond or, if the water is safe, a swimming pool, with plantings of

Water delivered by gravity (natural flow). Note protection of the spring; a shallow trench on the upper side to divert surface water, and the site inclosed by a fence to keep out stock. (U.S.D.A.)

local aquatic plants along the drainage area, flanked by rock gardens or shrubbery.

Springs should preferably be enclosed by a rectangle of masonry deep enough to exclude surface water or shallow soil flows and should be covered to keep out leaves and trash. Failing springs in springy hillside locations can often be rehabilitated by carefully digging out the natural basin and filling with coarse sand and gravel. In some cases, underground water moving downhill can be intercepted with lines of field drainage tile in the form of flanking wings which discharge into the spring basin.

In areas subject to freezing, water supply pipes from the spring to the farmstead buildings must be protected. Where trenching the pipe underground below freezing level is impossible due to rocks or stones, freezing can often be prevented by permitting a

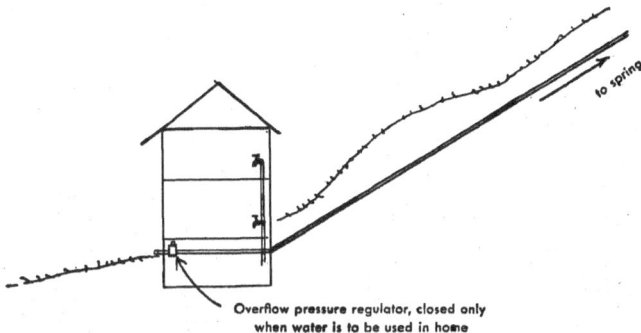

Overflow pressure regulator, closed only
when water is to be used in home

The constant-flow spring water supply system prevents freezing of pipes or stagnation of water in pipe system. The overflow pressure regulator, set to flow when faucets are closed, closes whenever water is drawn in the home.

constant flow of water through the pipe. Under such conditions, practically the whole pressure from the spring level may be utilized at the faucets and taps of the water supply system, if a pressure-regulated overflow valve at the end of the piping system is adjusted for slightly under the spring head pressure. (See sketch.)

In addition to furnishing a water supply, springs have for ages been utilized in connection with the cooling and storage of foods. Water temperature between 45° and 55° provides satisfactory cooling for dairy products immersed in a trough in the spring run. The tempering value of the water keeps the spring house cool in summer and tends to prevent freezing in winter. The humidifying and cooling effect of running spring water along the floor of the vegetable or fruit storage is highly beneficial. Fortunate indeed is the owner of a dependable spring of cool, clear, potable water; nostalgic the boyhood memory of crisp, cool watermelons chilled in an overnight spring bath.

Wells and Pumps

In the absence of springs, ground water reservoirs must be reached by wells. Wells are usually classed as deep or shallow according to whether they can be serviced by a suction pump effective for water lifts not to exceed 25 to 28 feet, or whether the pump cylinder or lift mechanism must be immersed in the water.

Shallow wells are often made by digging a round pit, 3 to 4 feet in diameter, to below the ground water table, and lining the pit with stone, block, or brick laid without mortar, except for the upper 6 feet and the foot or so extension above ground, which is cemented to exclude surface waters. Such wells are best capped with a concrete cover provided with a manhole for servicing and inspection. Provided the water is within vertical suction lift of the pump, pumping equipment can be located at a considerable distance from the well, preferably where accessible in a non-freezing basement room of the house. Where the water table is slightly below the lifting capacity of the pump, the well can be enlarged at the top to constitute a pump pit, thus bringing the pump within water-lifting distance and at the same time protecting it from freezing. In areas where underground strata make penetration possible, shallow wells are often driven instead of dug. The well pipe is capped with a special pointed driving head having a perforated section. This is driven by hammers or falling weights, with pipe sections added, until a water supply stratum is encountered. Where the flow to one point is meager, several such points may be driven at intervals of about 20 feet and all connected to the one suction line of the pump.

Deep wells are usually drilled. This requires special equipment and the job is usually handled by professional well drillers. As the hole is drilled, a steel tube, called a well casing, is driven down to avert earth cave-ins and to keep surface water from entering the completed well. In areas having underlying solid rock, the casing is driven only to make a tight fit with the rock. The usual practice calls for a 6-inch casing, since this will accommodate good-sized pumping equipment sufficient for most farm needs. Much can be done by an experienced well driller in developing the flow of a well and conditioning it to provide the maximum water supply. Local water-divining experts to the contrary, no scientific basis has yet been discovered for locating underground water supplies by the peach or willow twig method. Since deep well pump heads must be located directly over the well hole, it is desirable to locate the well near the house, so that an extension of the basement may serve as a pump room, giving all-weather access for maintenance and protection from freezing.

Pumps are literally the heart of the water supply system. They are roughly classed as deep-well and shallow-well. Shallow-well pumps are basically of the suction type, the pumping mechanism creating a partial vacuum in the pump into which the atmospheric pressure forces the water. The water thus forced up through the pipe to the pump is discharged under pump pressure to the point of use or to the storage unit. The most common type of shallow-well pump is the reciprocating or piston pump. Usually this is belted to an electric motor or gasoline engine, although the old-fashioned pitcher pump is a fine example of a hand-driven unit. Such pumps are easy to service, relatively low in upkeep, and maintain their efficiency over long periods of use.

Centrifugal and turbine type pumps are also used for shallow-well service. These are usually directly connected to a motor in standard water supply systems. Such units are particularly compact, require very little attention, are quiet in operation, are quite efficient within their pumping pressure range, but are harder to service. Of late years, the so-called jet pump has been refined and perfected and is increasing in popularity. This is essentially a shallow-well type pump which by-passes part of the outgoing stream for use in boosting the water in the well pipe to within suction limits. This makes it possible for the pump to be located at a distance from the well even though the water depth exceeds normal suction limits. Price ranges between that of the shallow-well and the deep-well pumps.

Deep-well pumps are located directly over the well hole, with part of the pump mechanism extending below the water level. The reciprocating plunger and cylinder type of pump is the most commonly used. A cylinder forms part of the well or drop pipe which extends from pump head to below lowest water level. A piston in the cylinder connected by a sucker rod to the pump mechanism lifts water out of the well. Check valves at the cylinder

11

or end of the pipe prevent run-back of water. This type of pump represents two separate units—the pumping head above ground and the plunger, cylinder, and pipe connection below. Capacity

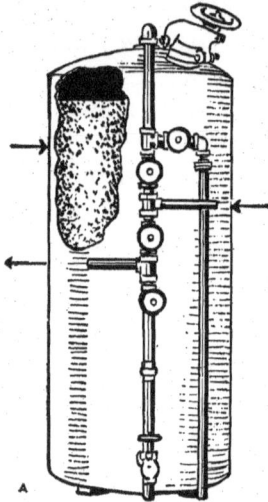

A.—*Water softener, single unit, consisting of zeolites in a steel tank.*
B.—*Deep-well motor-driven pump head. Must be located directly over well hole.*
C.—*Shallow-well pump, mounted over pressure tank, can be located away from well but preferably not more than 24 feet vertical distance above water level.*

depends on length of stroke and size of cylinder. A desirable form of this assembly, in which drop pipe and cylinder are the same diameter, permits withdrawal of plunger and valves without pulling up the drop pipe.

Where large volumes of water are required and for deep wells particularly, turbine type pumps are often used. In these a shaft extends from pump head to below lowest water level, encased and supported in a well pipe. A series of cup-shaped impellers are mounted at intervals on the shaft, which, at high shaft speed, propel the water from one lift unit to the next and out through the pump head. Modern pumps of this type are. usually directly connected to the shaft of a vertical motor, although belt-driven units are common. The force of compressed air is utilized in a third type of deep-well pump. In this, air from a compression tank is alternately admitted to a pair of cylinders below water level, displacing the valve-trapped water and forcing it up out of the well pipe.

Choice of the water supply system and particularly the pumping unit should take into consideration, as well as local conditions, the type of well, the volume of water supply and, above all, the maintenance factor. It might be more desirable to purchase from local sources maintaining a prompt repair service and a supply of replacement parts than to order a more perfectly engineered and constructed unit on which local service is unobtainable. This truism, incidentally, holds for practically all types of farm equipment.

Water Storage Systems

A complete water supply system includes delivery of the water to the point of use: Since it is obviously undesirable to have the pump start and stop for each opening of a faucet, some form of water storage between well and faucet is necessary. Two systems are in common use. In one, the water is pumped to a gravity storage tank located at a conveniently high point to feed all points of use by gravity. There are numerous commercial as well as homemade controls for stopping and starting the pump according to tank water level. Where the tank is considerably removed from the pump unit, such control becomes more difficult. Pressure switch controls for gravity tank systems are usually not sensitive enough, since in a tank located 40 feet above the level of the pump and having a vertical capacity range of 10 feet, with water pressure .43 of a pound per foot of elevation, there would be only 4.3 pounds differential pressure between a full and an empty tank. In gravity tank systems, considerable piping may be saved by using only one pipe connected to the bottom of the tank for both the filling and emptying cycles.

The hydro-pneumatic water supply system is usually considered most satisfactory. In this, the water is pumped into an enclosed tank, usually a horizontal or vertical cylinder, thus compressing the entrapped air and placing the water under air pressure by which it is forced to points of use. Such systems ordinarily operate between the limits of 20 to 40 pounds pressure. This differential lends itself well to automatic pressure-switch control for starting and stopping electric motor-driven units. The same standard switch can be used for stopping engine-driven pumps by breaking the ignition circuit. A combination of the hydro-pneumatic and gravity systems is often used on livestock and dairy farms. This has certain advantages in that the water for household usage is under hydro-pneumatic pressure and control, giving a quick flow through relatively small household pipes, while a gravity tank in the barns provides water for drinking cups, float-control troughs, and other points of use at desirable low pressures. Too, sticking of a cup valve or trough float will result in the loss of only one tankful of water instead of continuous operation of the automatic pump; tank water

Hydro-pneumatic Water System. Electrically-operated, fully automatic.

is only pumped against a pressure of 10 to 12 pounds instead of 20 to 40; tank location in the hay mow above the animals permits tempering of the water by animal warmth, thus tending to higher winter water consumption and increased animal production; storage of a day's supply to offset the possibility of pump repairs costs considerably less in open-tank capacity than in a pressure tank.

Water Piping

Piping between buildings on the farmstead is best underground, both for protection against frost and for greater sightliness. Subsequent removal for inspection and replacement can be greatly facilitated by threading the water pipe through a line of cemented bell-end vitrified tile leading from the house basement to the barns. This obviates the necessity of redigging the trench, disturbing the lawn, or thawing the frozen ground in case of a winter pipe-freeze. Since most water contains iron or minerals which are gradually precipitated in the pipe and because pipe-size is reduced through such rust accumulations, iron water pipes with a minimum inside diameter of an inch are best for supply line service. For the household water supply system, $\frac{1}{2}$- to $\frac{3}{4}$-inch flexible copper tube or drawn brass piping offers advantages in easy installation, long life, and maintenance of effective diameter. The development of self-soldering joints and compression couplings for the copper tubing and brass pipe respectively has materially reduced the labor cost of installation.

Fire Protection Precautions

An appreciable degree of fire protection can be attained by providing suitable hose connections to the main pipe line in each of the farmstead buildings. Where freezing of exposed lines or connections is likely, a frost-proof, self-draining hydrant should be provided. Coils of rubber hose or small-size cotton hose should be mounted on supports in strategic locations ready for instant use. Adjustable hose nozzles are recommended so that the type of spray can be fitted to the character of the fire. In most cases, a fine fog-like mist under considerable pressure is most effective in blanketing fires. Where water supply and fire protection systems are combined in this fashion, pumps should have a minimum capacity of several hundred gallons per hour, and the automatic control switch should be so arranged that it can be blocked out of operation or quickly reset for higher pressure. This is done in order to insure continuous pumping during the fire. Otherwise the switch might cut out at 40 pounds pressure and would not cut in again until the pressure had been reduced to 20—both pressures being too low to insure satisfactory fire-hose supply.

HOT WATER SUPPLY

Hot water on tap the year round is a modern household convenience, as desirable in the country home as in the city. Between the old-fashioned "water-back" heater on the kitchen range and the thermostatically controlled, insulated, electric or gas heater, there is a wide choice and variation of hot-water heating systems. Choice will necessarily be dictated by the form of heating energy available and possibly by the type of house heating system in use. The wood- or coal-fired cook stove with water-back heater transferred heat to water in the reservoir alongside the stove or to a hot-water tank located near. This is satisfactory only if the cook stove is operated rather continuously.

Bucket-a-Day Stove

Perhaps most generally used is the so-called bucket-a-day or hot-water pot stove. This is a small, cast iron, water-jacketed, top-feed stove available in various sizes measured in terms of the number of gallons of water it will heat an hour. It is connected to the so-called range boiler or hot-water tank by suitable piping so as to insure circulation and quick heat-transfer from stove to tank. The shorter the piping connection between the two, the less troublesome the system is likely to be. From the standpoint of heat-loss economy, however, the hot-water tank should be located rather centrally with reference to the different points of hot-water use, so as to minimize the waste of cooled water in long pipe lines between heater and faucet. To reduce the volume of water subject to pipe cooling, the hot-water distribution pipes should be as small as permissible to give the volume required. Copper tubing and brass pipe of $\frac{3}{8}$- and $\frac{1}{2}$-inch inside diameters are extensively used for hot-water piping because of their corrosion resistance and their maintenance of full diameter effectiveness.

For those who do not mind the twice-a-day chore of tending the fire, the bucket-a-day stove will prove very economical in operation costs and low in first installation cost. Daily requirements will range between 25 and 50 pounds of coal. With a good chimney draft, sizes as small as pea coal can be used, although larger sizes are more commonly recommended.

Furnace Heating for Hot Water

Steam and hot-water furnaces can readily be fitted for water heating. The simplest attachment is the ordinary furnace coil, consisting of a bent section of pipe extending into the fire box and connected to the hot-water tank circuit. Because of the intense heat and its localization, such coils quickly fill up with precipitated lime and iron from the water and must be replaced frequently. A slight improvement is the water-back heater, consisting of a cast iron shell fitted for pipe connections to the hot-water tank circuit. Either of these units can be used in all types of heating furnaces, although they are more commonly applied on the hot-air circuits. The regenerator coil heater commonly fitted to hot-water and steam furnaces consists of a cast iron jacket mounted outside of the furnace through which the furnace water circulates. Heat is transmitted to copper coils within the jacket, which are connected to the hot-water tank circulating system. This system is least likely to precipitate minerals and iron from the water being heated and, although it costs slightly more to install than the other types, it is considered well worth the difference in giving lasting satisfaction.

Both pot stove and furnace heater can be connected to the same hot-water tank circuit. A cut-off valve in the pipe leading to the furnace heater should be used to prevent the loss of pot stove heat into the furnace boiler during the idle season. Automatically fired furnaces of the oil burning, gas, or coal-stoker types are frequently used the year round for water heating. In such cases the furnace should be very well insulated and all circulating heating pipes to the house shut off, so that heat is confined to the boiler and used only for hot-water heating through the regenerator. Although the cost of operation is usually more than that required for direct heating of water, the convenience and automatic operation make the system attractive to those not caring to install supplementary pot stove heaters or automatic electric or gas water heaters. Since the latter are specifically designed for their job of hot-water heating, they represent the most trouble-free, efficient, and clean means of providing the luxury of hot water on tap at all times.

Hot-Water Insulation

Insulation of the hot-water storage tank is not only desirable but necessary from the standpoint of heat and fuel conservation. Standard insulation jackets are available from the trade, but homemade jackets can be fabricated that are equally effective and represent a saving for those who care to do the work. Such jackets can be made by wrapping the tank with mineral-wool insulation batts to a thickness of at least 4 inches, finishing with an outside wrap of heavy kraft paper or roofing paper held on with wire or straps. This, for appearance' sake, can be coated with aluminum paint. The exposed dome or top of the tank can be fitted with cut pieces of the batts or plastered to a depth of 4 inches with asbestos furnace cement. Insulation of the hot-water pipes leading from furnace to tap, especially in the exposed basement portion, is another worth-while economy, particularly in the case of iron pipes of larger diameter.

Many excellent bulletins by Experiment Stations are available on the subject of home plumbing and water supply. Similar information is provided through the United States Department of Agriculture, Bureau of Agricultural Chemistry and Engineering.

SEWAGE DISPOSAL AND PLUMBING

Present standards of living make modern house plumbing systems mandatory. However picturesque or camouflaged the little backyard shack so fabled in poetry and prose, however devious or landscaped the deep-worn pathway to its door, its' presence on today's farmstead is largely a matter of emergency convenience or an expression of individual eccentricity.

The Bathroom

A modern bathroom can be realized under practically all conditions, provided the farmhouse is equipped for running water. Space requirements are no hindrance, since a lavatory, toilet, and bath tub installation can, if necessary, be fitted into a room no larger than 5 by 6 feet. Cost of fixtures is no deterrent either, for strictly utilitarian units can be purchased at surprisingly low cost. There is nothing mysterious or difficult about the installation of plumbing fixtures and services, and many a modern bathroom is a reflection of the handy man's spare time. Care should be taken, however, that all work is performed in accordance with standards set by plumbing codes, in order that health is not endangered and that the job will give trouble-free satisfaction. As with many remodeling and modernizing jobs, those who find it possible to do so will probably be best satisfied by entrusting the work to skilled workmen. This usually insures a more satisfactory installation, may prevent costly and aggravating mistakes, and afford the owner the benefit of years of specialized experience. In plumbing installations particularly, careful planning can materially reduce the cost through a compact arrangement of units and piping.

Cesspools

Of extreme importance in the planning of house plumbing is the sewage disposal system. Architects and even plumbers frequently give this secondary thought. Wherever house sewerage is exposed on the ground surface a nuisance results, capable of endangering community health.

Cesspool disposal is too often recommended. In this, sewage is piped to a hole in the ground which is usually lined with dry-laid brick, stone, or block and provided with some form of cover. Liquids are expected to disburse through the open bottom and through the sides into the soil. It is obvious that soil conditions which favor uninterrupted flow from the cesspool also favor movement of sewage into wells of the surrounding territory. Where subsoils are such that a favorable filtering action results, 'movement is restricted and sewage solids and fats soon plaster sides and bottom of the cesspool, rendering it increasingly impervious to the outflow of liquids and eventually resulting in overflow. Cleaning out such reservoirs is an unpleasant and a decidedly temporary expedient. Treatment with a carboy of commercial grade sulphuric acid is generally effective in dissolving fats and reducing the solids, but no such treatment is of permanent value.

Septic Tanks

In view of these cesspool difficulties, the septic tank disposal system is highly preferable. Septic tanks may be built in place or may be portable special metal tanks available commercially. Although widely used, and giving fair satisfaction, commercial tanks of metal or concrete are considered by many less desirable than the cast-in-place, concrete, two-chamber or partitioned tank. Commercial units are often overrated as to capacity.

Since septic tanks are watertight, a separate disposal system for the liquid overflowing is provided. This consists of one or more lines of field tile or special disposal tile laid in trenches 20 to 30 inches deep and graded on a contour so as to give not more than 2 or 3 inches slope per 100 feet. Under most soil conditions, approximately 20 feet of this disposal system is provided for each person using it. Tanks should have a capacity sufficient to repre-

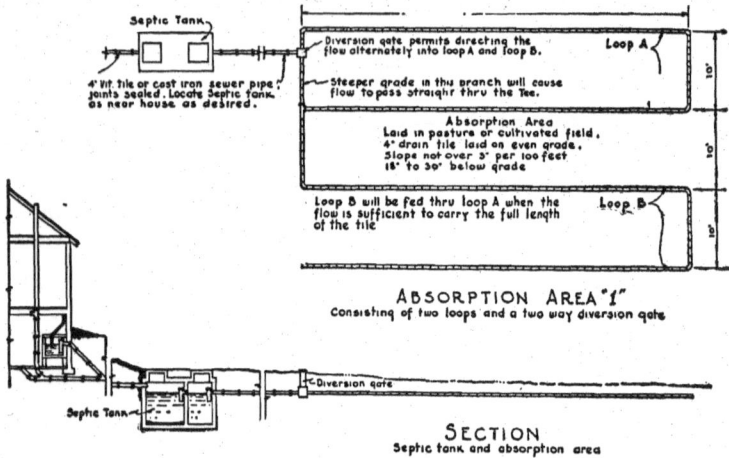

Plan and Section of an Entire Septic Tank Disposal System. Note grease trap, clean-out, and vent pipe.

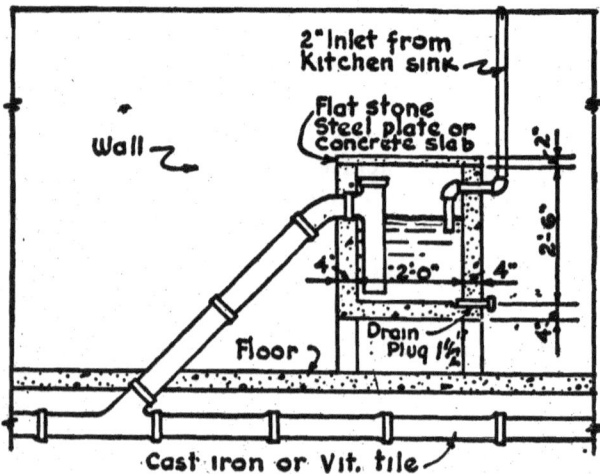

The Grease Trap of the septic tank disposal system shown above.

sent a 36- to 48-hour sewage flow. Usually 100 gallons of tank capacity is allotted for each person in the household. Tile lines are usually 4 inches in diameter. In any but the most porous type of soil, it is desirable to bed the tile in very coarse sand or medium gravel, filling the upper one-half to two-thirds trench depth remaining with surface soil. Specific instructions for the installation of a septic tank and the disposal system are available from most Experiment Stations and from the U. S. Department of Agriculture.

Suggested Sizes for Septic Tanks

| Number of Persons | Inside Dimensions | | | | Approximate Capacity | Over All Dimensions Width, Length, Depth |
| | Width | Depth of Water | Length | | | |
			First Chamber	Second Chamber		
	feet	feet	feet	feet	gallons	feet
4	3	4	3	2	450	4 × 6½×6½
6	3	4	4	3	630	4 × 8½×6½
8	3½	4	5	3	840	4½× 9½×6½
10	4	4	5	3	960	5 × 9½×6½
12	5	4	5	3	1200	6 × 9½×6½
15	6	4	5	3	1440	7 × 9½×6½
20	6	4½	6	4	2025	7 ×11½×7
30	6	6½	6	4	2925	7 ×11½×9

LIGHT AND POWER

Light was the first Creation. Ever since, man has concerned himself with the perfection of substitutes for natural lighting to dispel darkness. Fagot torch was replaced by fat and oil lamps; these in turn with wax and tallow candles; the discovery of petroleum soon spread the yellow glow of the kerosene lamp throughout the world. Then Edison unlocked the doors to a new pathway of progress leading today to the omnipresent electric incandescent and fluorescent lamps.

Few people consider their farmstead and home plans complete without including electric light and power service. Power transmission lines thread our highways and byways. Some states, such as New Jersey, can boast of nine out of ten farm homes with central-station electric service. The program of rural electrification fostered by power companies and governmental agencies is gradually reaching out to all but the most isolated homes. Even these need not be deprived of the advantage of electric light and power service. Private electric-generating plants in a wide variety of sizes and types are available. During periods of depression, electrification programs lagged; extension of service is halted by war-time needs of strategic materials; otherwise the history of electrification is that of an ever-expanding service.

There are three distinct types of private electric plant, each of which is specifically adapted to a particular class of farm service. The storage battery plant comprises a battery of wet storage cells and a generator operated by engine or water power. A 16-cell battery determines a 32-volt plant; 56-cell battery a 110-volt plant. Batteries are usually of the lead plate, glass jar type and last from 6 to 8 years in normal farm service or until precipitated sludge fills the jars to the bottom of the plate level. Nickel-iron batteries of the Edison type have less individual storage capacity and are considerably more expensive, but last longer and are less troublesome. With such equipment, energy is derived directly from the battery for lights and small power. Most plants are designed to start automatically for recharge 'when batteries become low. Operative efficiency is assured by the fact that the generator has a uniform battery-charging load, independent of service use. The second type of individual plant is equipped only with a small starting battery and designed to start and operate whenever a switch is thrown on the line. Such plants eliminate the large storage battery investment and are more efficient on full capacity loads, but prove rather expensive to operate when a few individual lights or a small appliance are used. Battery-type plant operating costs range between 2 and 3 times the average for central-station service; a direct-connection-type plant from $1\frac{1}{2}$ to 5 times that of the high-line service. Where large amounts of power are required and the diversity and constancy of use insures a rather high demand, the Diesel electric-generating unit has been developed to meet such requirements. Initial cost of such installation is, of course, relatively high, but operating costs per unit of energy are at most no more than high-line service.

When wiring the house and farmstead for private-plant electric service, the possibility should always be kept in mind that central-station service may at some time be available. The wiring should therefore meet pertinent electric wiring codes and in every way represent safe, serviceable, and satisfactory standards. Ability to get electric service should not be taken for granted. Homes have frequently been wired and equipment purchased in anticipation of service promised by over-enthusiastic salesmen. Power companies plan extensions according to their program of expansion, and it is always well to consult them regarding possibilities before making any commitments.

The planning of a wiring system for the house and farmstead should be a cooperative effort between owner and contractor. Neither is likely to have all of the information necessary to a satisfactory job. Recommended practices are detailed in many commercial and government bulletins and texts. The following suggestions are quite generally accepted. Single-phase three-wire service combining a choice of 110 and 220 volts is most common and satisfactory where separate motors do not exceed 7½ to 10 horsepower. To provide for the multiplicity of uses to which electricity is put in homes, nothing less than a 60 ampere entrance service should be used. Attempts to save on installation costs by skimping the number of convenience outlets is poor economy, since additions later will cost much more than they would have if originally included. The better plans call for a duplex convenience outlet for every 10 to 12 lineal feet of wall space and a minimum of one in each of the four walls of each room. To prevent stooping, outlets are often placed at the convenient height of 30 inches. Modern lighting systems emphasize portable table, floor, and wall lamps. Permanent center ceiling fixtures are passé except in kitchen, hallways, and over the dining room table. Meters are now generally installed in outdoor cabinets, obviating the necessity of entering the house to read them. A considerable saving in wiring costs can usually be effected by bringing the service wires to a load center common to farmstead buildings, with the meter and main power switch located on this pole and wires radiating out from it to the several points of use. Where sightliness is a considerable factor, wires may be run underground in non-metallic sheathed cable at only a slight increase in cost. Where the farmstead is located a considerable distance from the road, it is best also to locate the transformer at the load center. Most power company engineers recognize this advantage and will cooperate. This applies particularly to the heavier power installations.

HOUSE HEATING SYSTEMS

Our bodies are conditioned to a temperature environment ranging somewhere between 65 and 85 degrees Fahrenheit.

Deviations from this cause discomfort. To avoid this, in climates of occasionally lower temperatures, means must be provided for heating the house. Temperature and comfort, however, are not a parallel relationship, because humidity influences our reactions to temperature. Dry air at 70 degrees Fahrenheit may cause a sensation of chilliness by reason of rapid evaporation from the skin, while moist air at the same temperature feels comfortable. Cold walls or objects cause a feeling of discomfort by absorption of the infra-red rays emanating from our bodies; warm walls or those covered with insulation material reflect the rays and give the body a sense of warmth. Many such factors determine the choice and installation of a heating system, making it desirable to place the job in the hands of experts.

The oldest of modern heaters is the fireplace. Primarily of decorative or psychological value, it is effective in taking the chill from the room for off-season heating or as a booster unit. Notoriously inefficient, the standard fireplace has been greatly improved by incorporating an air circulation wall around the sides and back. This forces circulation of room air over hot fireplace surfaces, equalizing temperatures and more effectively utilizing fuel value.

Room heating stoves represent the next advancement. Still used where central heat is lacking, these offer a quick, inexpensive, and efficient heat service. They range from Franklin stove replicas to cabinet models simulating a piece of furniture and can be had for wood, coal, oil, or gas. When designed for forced air circulation, one such unit can serve several rooms and, under favorable circumstances, will heat a small house.

Central Heat

In the warm-air heating system, air is circulated over the furnace heating surface by either gravity or a fan and distributed to rooms through one large central register, as in the pipeless furnace, or by individual ducts direct to the rooms. Advantages of this system are quick heating, easy adaptation for air-conditioning both for summer cooling and winter humidification, and low installation cost, particularly in new construction. A piped warm-air system may be difficult to install in old houses with solidly constructed walls; heat control in individual rooms is occasionally difficult; hot-water heating is not so easily provided

:or as with radiator heating systems; and air ducts often act as voice tubes from room to room. Carefully designed and installed hot-air heating systems, however, combining air conditioning and forced circulation, offer one of the most modern and satisfying means of heat.

Steam and Hot-Water Heating Systems

In a steam heating system the furnace heat is utilized to convert water in the boiler to steam which is piped to radiators throughout the house. The one-pipe gravity system is the most commonly used. This has only one pipe connection to each radiator, the steam entering the radiator, condensing, and returning to the furnace by the same pipe. Displaced air is forced out automatic thermostat air vents. By fitting the system with special air vents, a partial vacuum system is created, resulting in much faster heating and more uniform room temperatures. Two-pipe systems for steam heat are an improvement, more expensive to install, but permit of a higher degree of flexibility. This is commonly called a vapor system.

Hot-water heating systems differ from steam in that the entire piping assembly including the radiators is filled with water which transmits furnace heat by circulation throughout the house. Because hot water is lower in temperature than steam, these systems require approximately one-third greater radiation surface than steam systems and are comparatively more expensive to install. Hot-water heat is particularly well adapted for colder climates, because of sustained warmth in the system between periods of furnace operation.

Combined Systems

Combinations of radiator and warm-air heating represent the latest development. In this, air is circulated over finned radiators by means of individual fans. This combines the advantages of warm-air-circulation heating with that of radiation heating and also offers a ready means for humidity control.

Insulation of pipes, ducts, and furnace is important to fuel conservation. Thin paper insulation or so-called insulating paints are practically useless on hot-air ducts and furnace pipes. Cellular asbestos pipe-covering or wrapped insulation is effective and worth while. Radiators and heating surfaces painted with metallic paints of the bronzing variety are less efficient than those covered with non-metallic paints. Most radiator grilles and covers inhibit heat diffusion.

Fuel

Wood, coal, coke, oil, or gas may be used in house heating systems. Where wood is plentiful and labor available for cutting, wood is entirely satisfactory as a furnace fuel. Hardwood varieties are preferred. Operation is limited to hand firing, more frequent attention is necessary, and a greater volume of ashes must be handled. These have fertilizing value for the garden, however.

Hand-fired coal systems constitute about two-thirds of the total central heating installations in our homes. Anthracite or hard coal and bituminous or soft coal are used according to proximity to respective coal-producing areas. Hard coal burns with a clean flame, has little ash, whereas bituminous is likely to be smoky and produces considerably more ash. Anthracite coal usually used for hand-fired boilers comes in five standard sizes: broken, egg, stove, chestnut, and pea, ranging from $4\frac{1}{2}$ inches down to $\frac{1}{2}$ inch in size. In general, the larger the fire box, the larger the size of coal which can be used. Large-size coals, however, are usually more costly per ton than the smaller sizes but, on the other hand, contain fewer impurities. Bituminous coal is less standardized, but the sizes corresponding to pea, chestnut, and stove in the anthracite scale are most commonly used. Anthracite lends itself best to the semi-automatic magazine-feed boiler and to stoker operation.

The modern coal stoker practically eliminates furnace drudgery. It takes coal from the bin, feeds it to the furnace as house heating demands, furnishes the required draft according to the rate of feed, and even removes the ashes to enclosed containers or separate ash pit. Stokers normally use buckwheat or rice coal. These range between 9/16 and 3/16 inches in size. These sizes usually cost materially less than the larger grades of coal, but because of the difficulty of removing foreign material have a higher percentage of impurities. Because of the coking and clinkering action of soft coals, fully automatic firing by stokers has not been entirely practical.

Coke is a popular fuel, particularly for hand firing, where available near industrial sections. Coal coke is a by-product of the production of manufactured gas from bituminous coal. Since the volatile matter has been removed in the process, it burns with almost no smoke, has little ash, and responds quickly to heat demands. Petroleum coke is a residue of petroleum distillation; it has characteristics similar to oven coke, but burns slightly faster. Coke for domestic heating is usually sold in sizes corresponding to the larger sizes of coal grades. Standard worm-feed stokers will not satisfactorily handle coke. Special type plunger-feed stokers for coke are in development.

Oil Burners

Oil burners have swept the country in popularity. This is because little attention is required for the operation; they are automatic and lend themselves readily to all types of heating systems. They can be installed in old boilers and furnaces or can be purchased as a part of a heating plant specially designed for oil burner use. Oil burner heat is intense, however, and it is essential that the furnace heat-absorption surface be sufficient to utilize the heat generated, in order that chimney losses and waste be prevented. Consideration should be given dependability of fuel supply, especially during winter months when roads may be blocked or when rail or boat transportation difficulties prevent shipments to particular areas. Oil burners depending on electrically operated fuel pumps and blowers and controls are vulnerable to interruptions of electric service during lightning, sleet, and wind storms.

There are two general types of oil burners, the gravity-flow vaporizing burner, utilizing a well or a hot plate, and the atomizing burners of either the rotary or the gun types. The hot-plate gravity-fed burners usually require a lighter, more costly grade of oil than the atomizing types. Maximum efficiency from oil burners can only be assured through carefully engineered installations and through periodic checks on valves, air, and control adjustment by competent service men. Next to gas, oil burners are considered the least troublesome, most satisfactory type of supply for home heating.

Gas Burners

Where natural gas is available or manufactured gas supplied under rate schedules designed for home heating, gas-fired furnaces represent the nearest approach to carefree heating. Gas-fired furnaces are usually designed for the purpose, although burners may be installed in heating plants designed for other fuel —often at some expense in efficiency. Utilities supplying the gas usually take the responsibility for adjustment of the heater to correspond to the particular grade or mixture of gas supplied. The so-called bottled or liquid gas is considered by most too costly for utilization in house heating.

Although traditionally located in the basement, the modern furnace and heating system can be so planned as to permit location in a room on the first floor level. This saves expense where excavation is difficult or ground water tables are high.

Fuel Conservation

Fuel conservation is of interest from the standpoint of individual and national economy. Those measures conducive to fuel saving often effect increased comfort.

Under the best of heating-plant efficiency, only a small part of the fuel value is actually retained for heating the home. The rest is lost through radiation from outside house surfaces, through the exchange of heated air with cold outdoor air, and through chimney losses.

Fuel burned for house heating can be conserved by (1) reducing air exchange between house and outdoors to a minimum, (2) lowering the rate of conduction and radiation through walls and roof, (3) maintaining room temperatures at a lower "comfort zone" level through humidity control and the use of low heat-absorption wall surfaces, and (4) efficient management and mechanical perfection of the heating system.

A fuel saving of 15 to 20 per cent can be expected by eliminating free air exchange between the house and outdoors. The following methods are suggested: Weatherstripping of all windows and doors; mastic caulking of open joints between the wall and the door and window frames; closure by caulking or stripping of all open joints between the foundation wall and the house sill; the sealing of all cracks in foundation walls and around the windows.

Heat lost through glass and doors is appreciable. The use of storm windows and doors and provision of enclosed entryways usually effect a fuel saving of 20 to 30 per cent. It is essential that door and window storm frames fit absolutely tight in order to prevent air exchange and loss of heat. A felt strip seal around the frame is sometimes used to give a tighter joint. Storm windows are particularly desirable in severe winter climates, since the direct heat loss through glass areas in the average dwelling is equal approximately to that lost through the remaining wall areas.

Insulation of the walls and either the roof or the ceiling of the house can be expected to effect a saving of 30 to 40 per cent of the fuel bill. One inch of insulation plus weatherstripping will approximate a 50 per cent saving, or insulation with storm windows alone as much as a 60 per cent saving. Wide variations in the aggregate value of these combinations of protective measures can be expected according to type of house construction, type and placement of insulation material, quality of workmanship, and range of temperature differential between house and outdoors. If only partial insulation is feasible, the attic ceiling or roof insulation is relatively the most effective for money expended. Structural insulation board is effective for surfacing rooms, for the attic floor, or against the roof rafters. So used, it converts an attic into usable room space for playroom, spare bedroom, or more orderly storage. If the studding space in side walls is not filled with insulation, all openings from basement to attic into this space should be closed with tight-fitting boards or stuffed with non-inflammable insulation batts. This has the added advantage of serving as a fire stop.

Absorption of radiator heat by adjacent outside walls can be reduced materially by placing a shield of metal or metallic-surfaced paper on a strip frame between radiator and wall and touching neither. This will reflect the heat back into the room and reduce absorption by wall area. Walls of plaster or tile give rooms a sense of chilliness because of the rapid absorption by these surfaces of infra-red rays emitted by our bodies. Backing the tile with insulation board greatly reduces this effect, as does the covering of plaster walls with paper or other heavy decorative surfaces.

Humidity

The air of our homes in winter is usually dryer than that of the Sahara Desert. This necessitates a high temperature level for a sense of comfort, due to the cooling effect of rapid evaporation from our bodies. By keeping the humidity in the home between 30 and 40 per cent relative, the temperature can be reduced to below 70 degrees and the feeling of comfort still maintained. Storm windows are also desirable in this connection to prevent condensation on the inside glass. Insulation of the outer walls may also be necessary to prevent condensation on these surfaces during extremely cold periods. Ordinary radiator pans, small furnace reservoirs, and similar devices for adding moisture to the air are relatively useless, because of the limited evaporative surface and the low temperature of the water. Devices which mechanically spray a fine mist of water into the air are much more effective. A homemade humidifier consisting of a wick of toweling or muslin extending from a water pan between radiator and wall on a wire frame across the top of the radiator and covered with a standard shield is very effective. From 3 to 5 gallons of water per day is necessary in a house of reasonably tight construction in order to maintain the desired humidity level. Hot-air heating systems lend themselves best to the addition of air moisture by the use of evaporator pans placed directly over the furnace unit or by forcing the heated air through automatically moistened pads located in the delivery pipes.

Livestock Buildings

Dairy Barns

Construction of dairy barns has quite generally become standardized throughout the country. Cows require protection from the elements, freedom from extremes in heat or cold, ventilation and sunlight, and sufficient floor area for comfort. Practically any structure which meets these requirements will represent satisfactory housing conditions. Climate will dictate the degree of enclosure and insulation. Although local practice and regulations of governing agencies will serve as a guide in planning, dairymen are quite agreed that a barn should be large enough to allow 500 cubic feet of barn capacity per cow; from 2½ to 3 square feet of window area per cow; 60 cubic feet of air change per minute per cow; and a preferred temperature range not exceeding the 40 degree minimum to 80 degree maximum spread from winter to summer. With regard to temperature, sudden variations are more undesirable than gradually attained extremes.

Small herds up to 15 milking cows can be arranged in a single row, preferably along the south side of an east-and-west barn. Where there is any possibility that larger herds are to be accommodated, standard barn construction providing two parallel rows of animals is preferred. In order to give the animals an equal

DIAGRAM SHOWING METHOD OF MEASURING
COWS TO DETERMINE LENGTH OF PLATFORM

USUAL DIMENSIONS FOR COW STALLS

BREED	*WIDTH—W	LENGTH M = L + 6"		
		SMALL	MEDIUM	LARGE
Holstein	3' − 6" to 4' − 0"	4' − 10"	5' − 2"	5' − 8"
Shorthorn	3' − 6" to 4' − 0"	4' − 8"	5' − 0"	5' − 6"
Ayrshire	3' − 6" to 3' − 8"	4' − 6"	4' − 10"	5' − 4"
Guernsey	3' − 4" to 3' − 8"	4' − 6"	4' − 10"	5' − 2"
Jersey	3' − 4" to 3' − 6"	4' − 4"	4' − 8"	5' − 0"
Heifers	2' − 9" to 3' − 2"	3' − 8"	3' − 10"	4' − 2"

Length of partitions S   3'—6" for cows 3'—2" for heifers

*No stall less than 3'—4" wide should be used for cows in milk

Cow Stall Dimensions in Dairy Barn (U.S.D.A.)

chance for direct sunlight, such barns are usually constructed in a north and south direction. Barn widths vary between 32 and 38 feet, with considerable preference shown for the 36-foot outside width dimension. Barn length is determined by the number of animals, allowing from 3½ to 4 feet of width for each stall along the length of the barn. To this must be added thickness of the end walls and space required for cross alleys. Dairy barns usually have two stories to provide storage room for hay and straw above the stable portion. Such storage may present a fire hazard, and the one-story milking barn for the dairy herd has therefore come into prominence. This necessitates separate storage of feed and straw and generally involves a higher investment.

Whether cows should face in or out is a controversial question. From the standpoint of barn cleanliness, ease of litter removal, arrangement of a common entry or exit to pasture, and convenience in milking, the out-facing arrangement is preferable. One of the most convenient standard layouts for barns up to 100 feet in length is to provide one centrally located cross alley terminating in the milk house on one side of the barn and the attached feed room and silos at the other side. This eliminates the necessity of cross alleys at the ends of the barn, thus saving this space for four additional cows. The last stanchion rail is set 12 inches from the wall to provide passageway for a man and reduce cow-heat absorption by the wall. With this arrangement, feed and litter alleys intersect at only one point, thus reducing disease possibilities. This arrangement has another advantage over placement of the silos at the end of the barn, in that expansion is feasible at either end.

Practically all construction materials are adaptable for dairy barns. Glazed tile, building tile with cement mortar or plaster finish, brick, concrete and cinder block, wood frame and sheathing construction, as well as wood frame encased in cement asbestos and metal sheathing—all are represented in the materials category. Consideration of the fire hazard has influenced many dairymen to build of fireproof or fire-resistant materials. A fire-resistant type of barn recently developed has many advantages. This consists of standard wood-frame construction, sheathed on the outside with corrugated or flat sheets of cement asbestos, on the inside walls and ceiling with plywood against the studding and joists, followed by a well-lapped layer of vapor-resistant paper and a finish surface of a thin smooth sheet of cement asbestos. The inter-stud and joist space is insulated with fire-resistant or mineral-wool fill insulation or batts. Such construction eliminates the paint schedule, permits thorough washing, presents a light, airy, smooth, unobstructed finish of maximum cleanliness. The same type of construction can be paralleled with the use of metal sheets to equal advantage, except that an occasional paint application will be required.

Ceiling insulation in barns should at least equal that of the walls in order that moisture condensation, if any, will favor walls instead of ceiling.

Ventilation is provided by using the gravity flue system or by means of electric fans. Windows do not constitute a ventilating system. Intakes should be spaced to give good air distribution and baffled so as to prevent direct drafts. Commercial units are fitted to close automatically against high winds. In the gravity system, one outtake flue is sufficient, if of proper size and construction, for a barn up to 80 or 100 feet long. The number of fan systems depends on their individual capacity, allowing 60 cubic feet of air change per cow per minute. Location is not critical. Commercial roof ventilators should always be connected down to the stable by means of an insulated duct. Those not so connected are largely ornamental in value.

Although the commonly recommended practice is for manure removal from the barn to the fields daily, snow and mud may prevent this schedule and some form of manure storage at least 50 feet removed from the barn may be necessary. To prevent the loss of nutrients by leaching and run-off, such a storage should preferably have a concrete bottom and side walls. A roof is desirable but often omitted.

With larger herds it is considered good practice to provide separate quarters for the appreciable number of young stock and dry animals. Many Boards of Health also insist that the milking herd be segregated.

Records show that cows having access to water at all times produce as much as 5 per cent more milk than when watered by hand or at a trough on a 2- or 3-times-a-day schedule. Drinking cups are, therefore, indicated, the individual type being the most common.

When wiring the dairy barn, it is well to plan the circuit so that rows of light over each feed alley section and over the litter alleys are individually switch-controlled. Over the feed alley one 40-watt lamp for each 4 cows is considered sufficient. Litter alley lights are spaced approximately 10½ feet apart.

Two types of mangers are commonly used. One, the sweep-in type, is formed by having a high feed alley on a level with the front of the manger (see Manger B on opposite page); the other, the raised-front manger, extends from 20 to 30 inches above the floor level (see Manger A on opposite page). Proponents of the sweep-in type claim labor saving as a prime advantage. Hay and feed can be kicked back without forking. Those favoring the raised-front type claim that guards will practically prevent throw-out of feed, that possibility of disease is reduced, that cows will not injure their knees reaching for feed they cannot see, and that

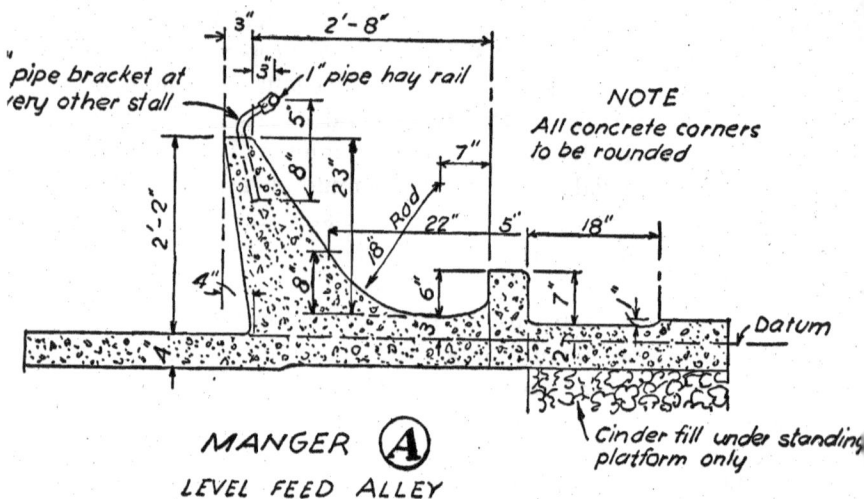

NOTE
All concrete corners to be rounded

MANGER **A**
LEVEL FEED ALLEY

Cinder fill under standing platform only

If manger divisions are to be used, manger curve should be formed with templet furnished by manufacturer

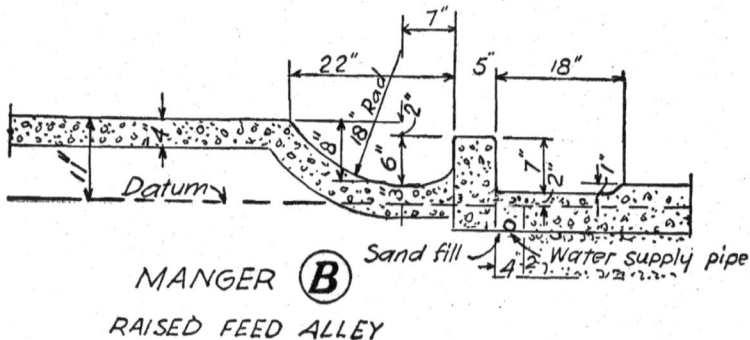

MANGER **B**
RAISED FEED ALLEY

Sand fill — Water supply pipe

this construction makes possible placement of feed alleys and litter alley on the same level, thus making ramps unnecessary.

The cost of constructing dairy barns varies widely in different sections according to completeness of structure, local building material costs, and labor. In the dairy belt, fully equipped barns usually run between $125 and $150 per cow capacity.

Poultry Houses

From a practical viewpoint, any shelter which protects the birds from the elements, from drafts, and from sudden variations in temperature and, at the same time, admits sunlight and provides ventilation will be entirely satisfactory for housing the poultry flock. That there are many interpretations of these requirements is evidenced in the 200-odd poultry-house plans listed by colleges and experiment stations in the United States. Where winters are severe, poultry-house ceiling heights should be restricted to reduce cubic feet of air space per bird, in order to confine the small amount of heat which chickens generate. Poultry houses, therefore, often have a shed type of roof, sloping from front to back, giving a maximum rear wall height of 5 feet and a front wall of 8 to 10 feet, depending on the depth of house. Houses usually face south, running east and west. For these, the depth varies between 20 and 30 feet. In narrower houses, the birds are more subject to drafts and chilling. Adequate lighting of deeper houses is difficult. Houses are usually divided into pens by solid partitions extending from $\frac{1}{2}$ to $\frac{2}{3}$ of the depth from front to back, the remainder being enclosed with wire mesh. Pens are customarily as long as the house is deep. Any number of pens can be built together to form a long house, accommodating the required number of birds. For economy in construction, advantageousness of window locations, and simplicity of design, popularity of the shed-roof house is deserved.

Where building space is limited and birds are confined to their pens during the producing lifetime, multi-storied houses are favored, usually of 2 stories, but houses of 3 and 4 stories are also common.

Most poultry experts are agreed that a minimum of 3 square feet of floor space per bird should be allotted; 4 square feet is preferable. If birds are crowded, disease and abnormalities naturally reduce the flock to about this level anyway. Approximately 1 cubic foot of air exchange per bird per minute is the ventilation index. This is secured by vertical flue-ventilation systems or by a circulating under-rafter system. One outtake flue per pen is sufficient if it provides $\frac{1}{2}$ square inch of cross section area per square foot of floor space. Air is usually admitted through muslin-covered windows or window slots. The under-rafter system admits air through the front wall at window-sill level by a baffle arrangement which, after circulation, follows the ceiling line out the front of the house between joists or rafters through an adjustable opening.

Poultrymen have a wide choice of roost location and arrangement. Quite standard is placement along the rear of the house at approximately the 40-inch level. Roosts are spaced 1 foot apart and should provide from 7½ to 10 inches of length per bird. A board platform to catch the droppings is placed 8 to 10 inches below the roost level, with a wire mesh screen protection directly under the roosts and along the front edges of the droppings platforms to prevent bird-access to the intervening space. A variation of this is low placement of roosts directly over a droppings pit on the floor. This requires less frequent attention and is preferred by many poultrymen as affording the birds greater utilization of pen area.

The question of type of floor is a controversial one. Standard poultry-house construction has long dictated wood floors on a raised foundation. This invites draftiness, floor condensation, and rat and rodent problems, with the result that concrete construction of the first or ground floor is rapidly gaining favor. If the poultry house is located on a well-drained site, preferably a southern exposure, and the floor has a sub-grade of gravel, sand, or well-tamped cinders, a concrete floor offers several advantages in warmth and lasting qualities. Foundation walls should, however, extend approximately 18 to 20 inches underground, with a projecting lip to discourage rats from burrowing underneath. A dry floor depends more on good ventilation and the building up of an insulating layer of heavy litter early in the season than on preventing ground moisture from working up through the concrete. The latter is an unlikely condition where drainage is good.

Insulation of poultry houses in cold climates is desirable to prevent freezing of combs and consequent loss in production and as an aid in ventilation. Ceiling or roof insulation is most important. This is usually provided by structural insulation board ⅝ to 1 inch in thickness. Wall insulation of closed-front houses is desirable and recommended. Insulation board can be used here also, but inter-stud fill-insulation of commercial materials or sawdust or shavings is frequently used. Some form of sheathing inside the studs is used to hold the material in place. Chickens will

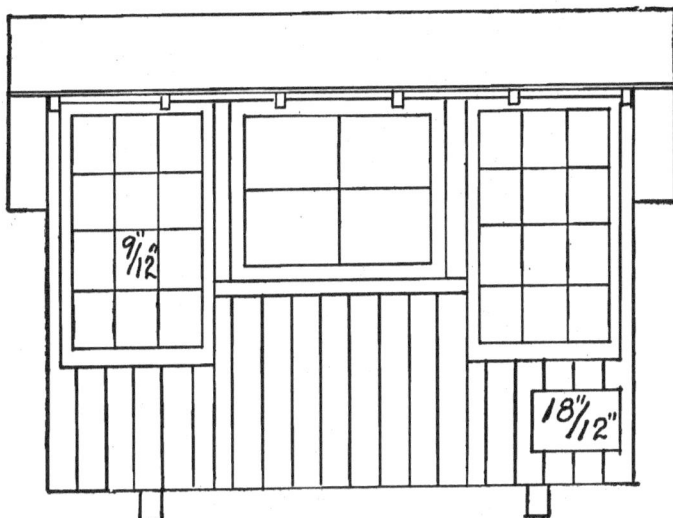

Front View, Side Windows Hinge in, Middle Window Slides Down Into Pocket

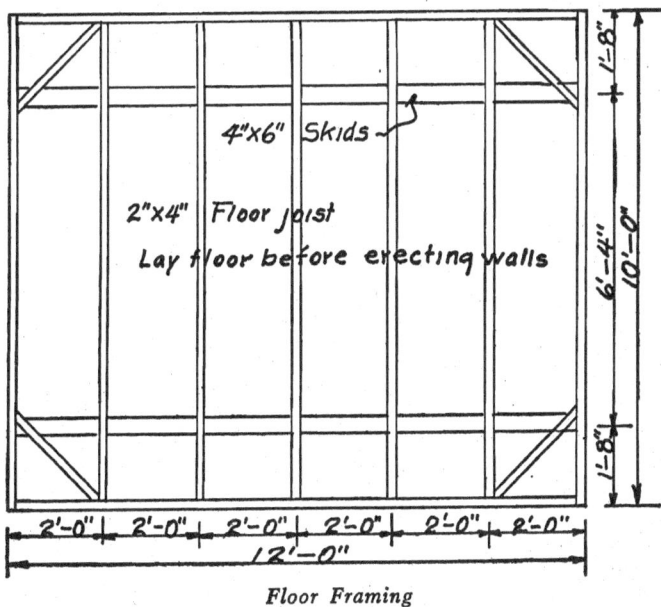

4"x6" Skids

2"x4" Floor joist

Lay floor before erecting walls

2'-0" | 2'-0" | 2'-0" | 2'-0" | 2'-0" | 2'-0"

12'-0"

Floor Framing

37

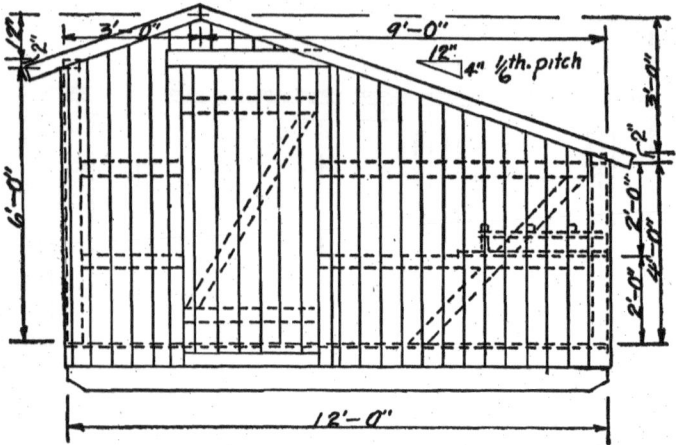

Side View, Showing Framing, Roosts, and Dropping Boards

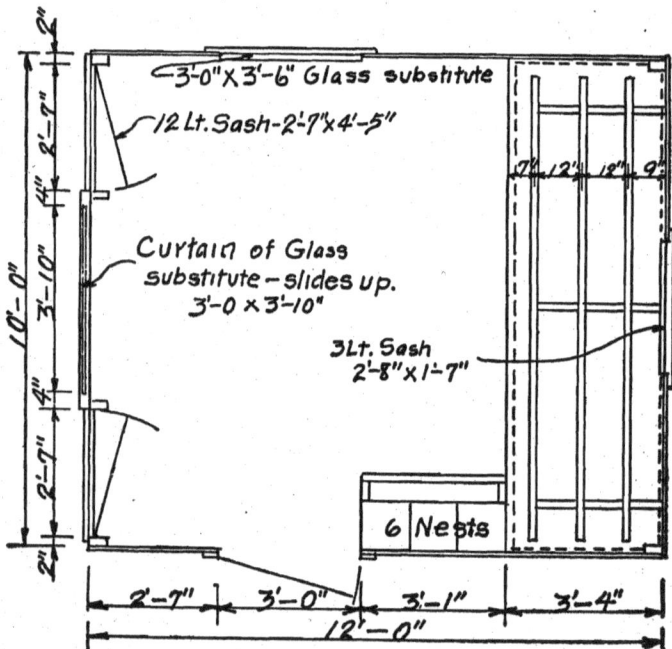

Floor Plan. Roosts and Nests Will Be Left Out When Used for Brooding

often pick and destroy ordinary insulation board within their reach. Hard surface paints, wire screening, plywood, pressed board, or a smooth plaster finish is used to prevent picking.

Maximum egg production is impossible without a constant supply of drinking water for the chickens. Many types of commercial and homemade fountains have been developed. Where the water supply is not critical, the constant drip fountain has some advantages. Water in motion does not freeze readily; birds drink directly from the drip or from small overflow pans, giving

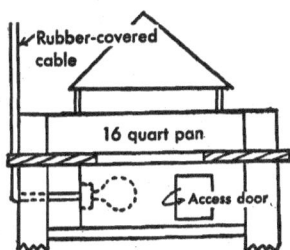

Electric light replacing oil lamp in poultry fountain warmer. (Use porcelain receptacle, closed terminals type.)

them cleaner conditions; less work is required for maintenance. Another form of piped supply fountain is float-regulated. This is economical of water use, but more vulnerable to freezing at night when birds are not drinking; it has proven very popular. Independent fountains are hand-filled and often have water-heating equipment incorporated in construction. It is important to prevent spillage of water on litter around the fountains, since this is a prime cause of excessive moisture in poultry houses during cold, muggy periods. Many poultrymen install floor drains connected to the fountain overflow. This not only helps prevent wetting of litter but greatly facilitates the daily cleaning out of fountain.

Poultry house lighting is necessary to encourage high egg production during the winter months. If egg prices during this period justify it, the use of artificial lighting on laying flocks will materially increase the number of eggs laid during the short day period. For simplicity of operation and control, the morning lighting method is preferred. Clock-controlled switches turn on the lights in the morning early enough to give the birds approximately a 12-hour day. Usually the poultryman turns the lights off when morning chores begin. An ordinary key-winder alarm clock can be used to turn the toggle-type switch. The evening lunch

Wires

Alarm winder key

Toggle switch

Alarm clock

Brace

3"

A

B

5" 4"

2 PIECES CUT TO FORM BRACES

6'6"

6"

SIDE ELEVATION

¼"   6½"   ¼"

A   B

7"

FRONT ELEVATION

CLOCK FASTENER

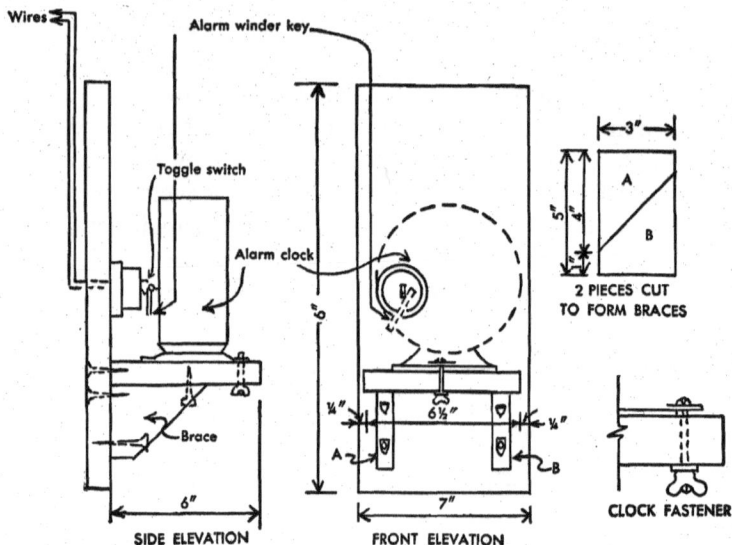

Clock Control for Poultry House Lighting

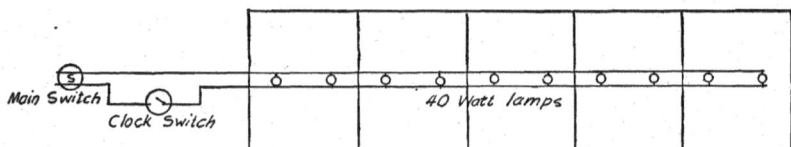

Main Switch

Clock Switch

40 Watt lamps

Simple clock control, particularly for morning lighting

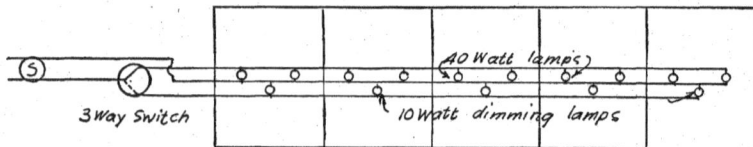

3 Way Switch

40 Watt lamps

10 Watt dimming lamps

Lighting circuit, using separate line for dimming

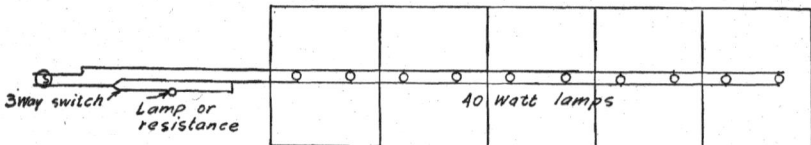

3 Way switch

Lamp or resistance

40 Watt lamps

Lighting circuit using lamp or resistance in series for dimming

WIRING DIAGRAM FOR THREE METHODS OF LIGHTING CONTROL

40

lighting period is also popular. With this, the day is extended by the necessary hours of artificial lighting or a 2-hour lighting period given the birds later on in the evening at a special feeding. This necessitates equipment which will turn the lights off as well as on. So that the birds may find their way to the roosts following the lunch hour, some poultrymen claim the necessity of

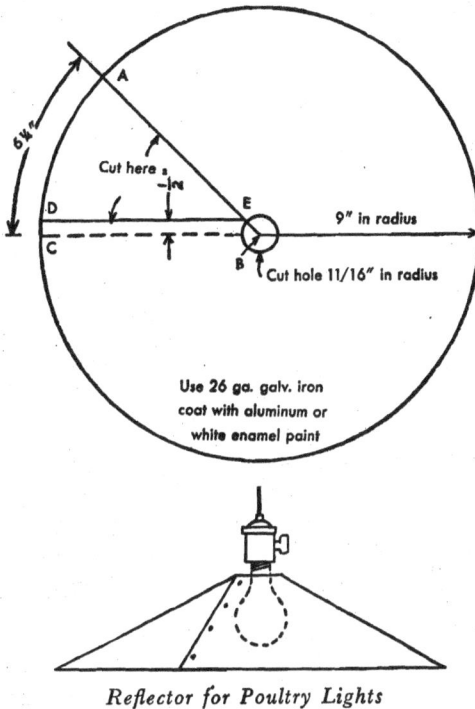

Reflector for Poultry Lights

a dimmer circuit or a separate lighting circuit with low-wattage lamps. During recent years, the all-night lighting practice with low-intensity lights has seen some favor. Instead of the usual 40-watt lamp for every 200 square feet of floor space, such lighting calls for just enough illumination for the birds to find their way about and feed. Lights are usually located halfway between the front of the house and the roost line and should be provided with reflectors to conserve light otherwise lost by absorption on ceiling and side walls. The lights should be lowered from the ceiling so that the light cones from the reflectors just hit the birds on the roosts.

When wiring poultry houses, it is quite desirable to run a 3-wire circuit. This will give two separately controlled 110-volt

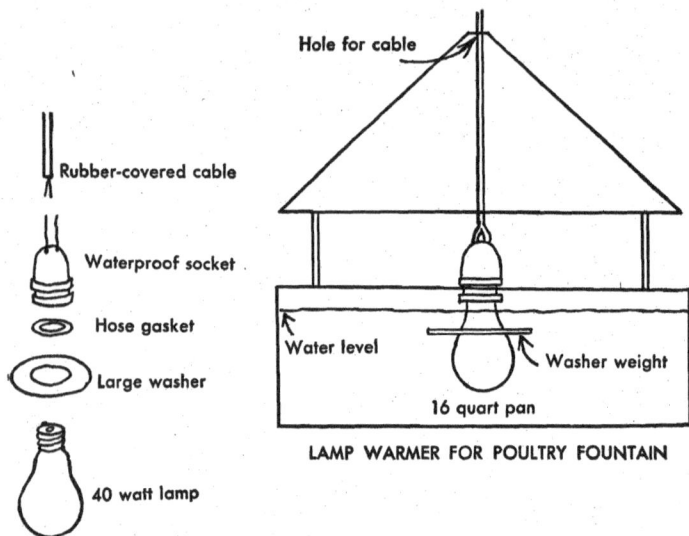

Hole for cable

Rubber-covered cable

Waterproof socket

Hose gasket

Large washer

Water level

Washer weight

16 quart pan

LAMP WARMER FOR POULTRY FOUNTAIN

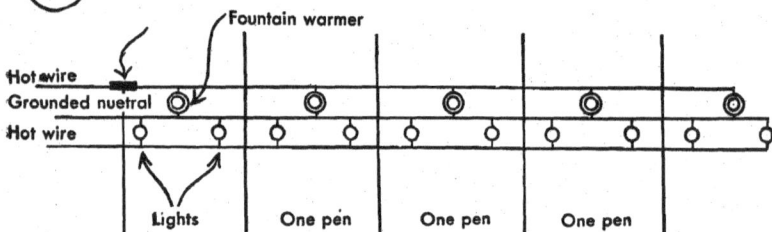

40 watt lamp

Fountain warmer

Hot wire
Grounded nuetral
Hot wire

Lights | One pen | One pen | One pen

WIRING SCHEME FOR POULTRY-HOUSE WATER HEATERS AND LIGHTS ON A THREE-WIRE CIRCUIT

(Wires should be soldered, wrapped with splicing compound, and covered with electricians' tape. Note arrow pointing to thermostat on the upper hot wire.)

circuits, one for lights and one for drinking fountain warmers, or for auxiliary and special lighting in particular pens. If some pens may be used for electric brooding, this type of wiring can supply the desirable 220-volt service for the electric brooders.

Hog Houses

Popular impressions to the contrary, hogs are normally a clean animal and prefer clean surroundings. Cleanliness pays dividends in more efficient meat production, reduction of disease, lowered

PORTABLE HOG HOUSE WITH SHED ROOF

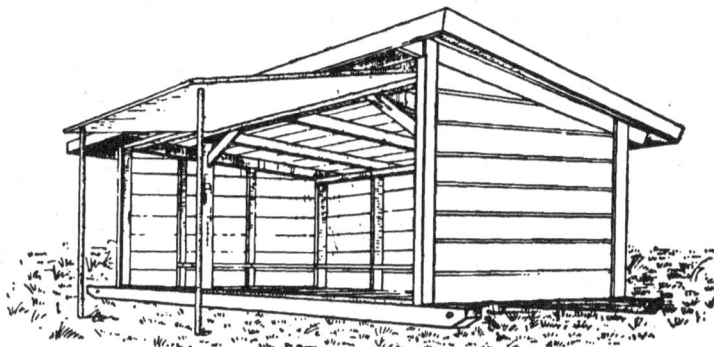

Perspective of Hog House with hinged door raised on poles to form a porch.

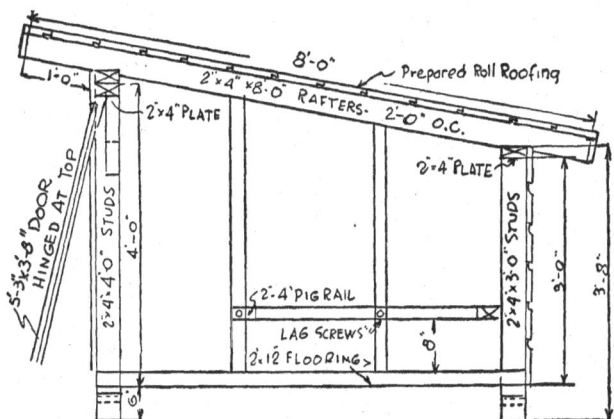

° SECTION °

BILL OF MATERIALS ·

Roofing	I Roll, 3ply prepared roofing
Hinges	I Pr. 6" strap hinges
Log Screws	7 - ½" x 8" Log Screws
nails	2 lb. 20d common nails
	2 lb. 8d Boxing nails.
	I lb. 6d Boxing nails.

PORTABLE HOG HOUSE WITH SHED ROOF

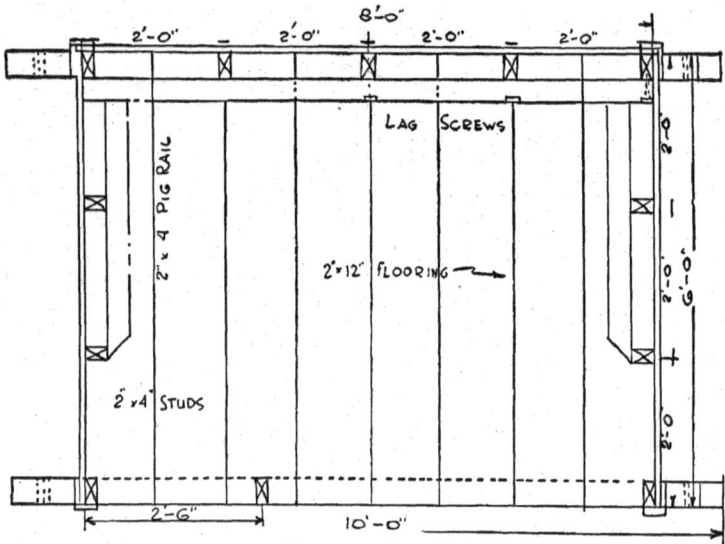

2'-0" 2'-0" 8'-0" 2'-0" 2'-0"

LAG SCREWS

2"x4" PIG RAIL

2"x12" FLOORING

2"x4" STUDS

2'-6" 10'-0"

• FLOOR PLAN •

2"x4" STUDS

2'-3"x 3'-10" OPEN ENTRANCE

5'-3"-3'-8" DOOR HINGED AT TOP

1"-4" STUDS

LAG SCREWS

PIG RAIL

1'-0"
6"x 6"

2"x12" FLOORING

2-4"x 6"x 10'-0" SKIDS

4"

3'-6" 6'-6"

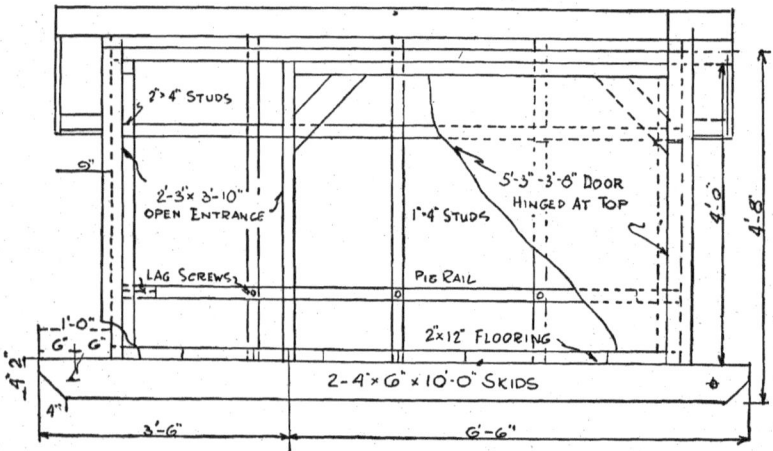

• FRONT ELEVATION •

Skids	2-4"x6"x 10'0"	
Plates	3-2"x4"x 8'-0"	
Rafters	5-2"x4"x 8'-0"	
Studs	3-2"x4"x 16'-0"	
Pig Rail	1-2"x4"x16'-0"	
Flooring	4-2"x12"x12'-0"	

Trim	2-1"x4"x 10'-0"
	3-1"x4"x16'-0"
Siding	30 bd. ft. 1"x6" novelty siding
Roofers	13-1"x 8"x10'-0"
Door	24 bd. ft. 1"x 6" novelty siding
	1- 1"x4"x 8'-0"

44

mortality. Best results are obtained where animals have a field run and are housed in portable structures that can be moved onto clean ground each year. One of the most popular hog houses is the individual or A-shaped house. This is readily portable, can easily be built with farm labor, costs little, is large enough to accommodate a sow and litter at farrowing time, is warm because of low cubical content, and can easily be cleaned by tipping it over for scrubbing and disinfecting. The house illustrated provides approximately 100 cubic feet of content and approximately 48 square feet of floor area.

To provide shelter for the growing litter, some hog raisers prefer the type of portable structure illustrated on page 296. This is a combination utility house which serves as a farrowing house and also a summer shelter by raising the drop door on posts imbedded in the ground.

Both the A-shaped and the shed-roof portable houses can be used for winter quarters by arranging them in a row facing the south or in a south- and east-exposure ell. For greater protection, straw can be packed between and at the back of the houses and held in place with a woven wire fence.

Permanent hog houses are preferred by some hog raisers. Such houses usually have a double or single row of pens the length of the house paralleled by a 4-foot work alley. Houses usually face south and, in most cases, have roof skylights to supplement the direct sunlight entering the relatively low side wall. Pens should preferably have concrete floors sloped to divert the liquids either to the outside of the house or to a separate gutter outside of the pen walk line. When used as farrowing pens, the floors are usually fitted with a wooden platform and pig guard rails. For the sake of warmth, houses should be constructed with minimum cubic content and with the ceiling or roof area insulated. Individual pens should approximate 50 square feet of floor area with a minimum width of 6 feet.

Type of house and arrangement of windows particularly will depend considerably on latitude and general climatic conditions, therefore making it desirable to check with local practice.

Horse Barns

Work horses are usually housed in the general-purpose barn with the other livestock to facilitate feeding and care and to share in the warmth. Many milk ordinances demand a solid partition between the milking herd and the horse stable. Work horses are usually kept in stalls, haltered to mangers. To prevent kicking and biting between horses, stalls are solidly partitioned to a 4- or 5-foot level and finished to the ceiling with a steel bar grating or heavy mesh wire. Comfort of the animal and its care necessitates a stall approximately 5 feet in width and about 7 feet in over-all length exclusive of manger. Fade-away or sloping-bottom gutters are often used in place of those with a rectangular cross section common to the dairy barn. Horse stalls are often floored with wood plank fastened to nailing strips imbedded crosswise of the stall in the concrete sub-floor. Floor boards should preferably be pressure-treated with creosote or wood preservative salt. Since almost 5,000 cubic feet of air per horse per hour is considered necessary for effective ventilation, either a gravity-flue ventilation system having at least 1 square foot of cross-sectional area in the outtake flue for each 5 head of horses or a ventilating fan of necessary capacity should be provided. The area of intake ducts should always exceed that of the outtake duct by at least 10 per cent.

Approximately 4 square feet of window sash area is desirable for each horse housed. Standard 4-light barn sash of 9 by 12 inches are recommended. Sash are usually framed to slide up between the stud frames, or they can be set to tilt in at the top. Window sills should be at least 4 feet from the floor, and it may be necessary to screen or bar the windows to prevent breakage by the animals.

Riding horses are usually stabled in box stalls. Such stalls preferably flank the south side of the barn, with individual pen doors to the outside. These are of the halved or Dutch type, permitting the top to open while the bottom is closed. Each stall is usually fitted with a metal-bar hay rack, a corner water bowl, and a manger or box for feed concentrates. Solid partitions should extend approximately 5 feet high and are continued to the required 7 foot height by the use of steel bars, gratings, or heavy wire. Box stall floors are surfaced with a mixture of clay and sand well compacted and puddled so as to form a hard but resilient surface. In some cases, end-grain wood blocks are set in asphalt on a concrete sub-floor. Such wood should be rot-

HORSE STALL

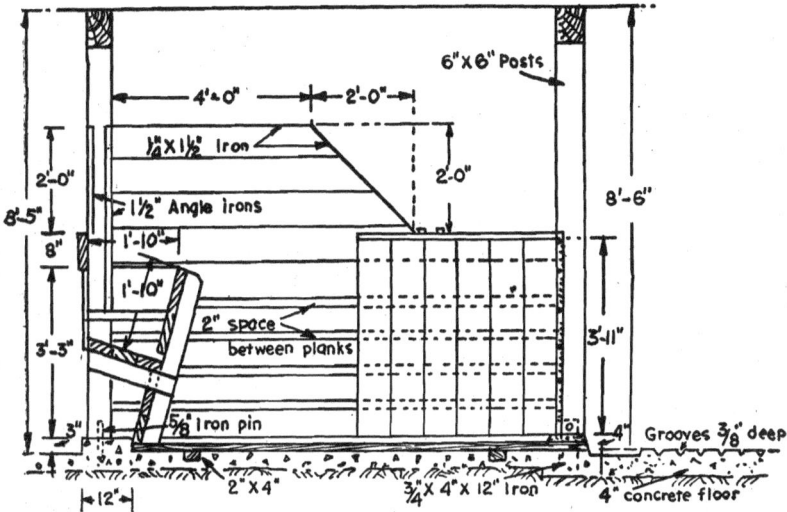

ELEVATION

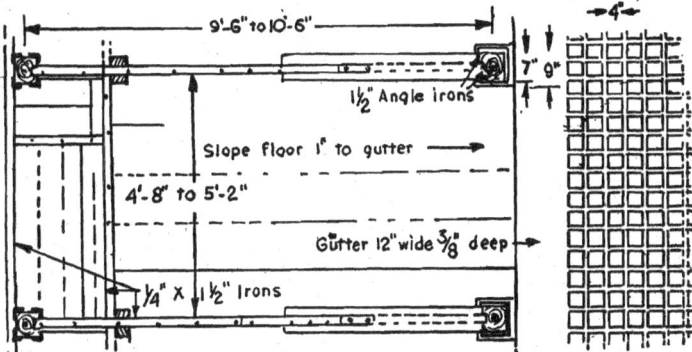

PLAN

(See opposite page for perspective and Detail "A")

47

HORSE STALL

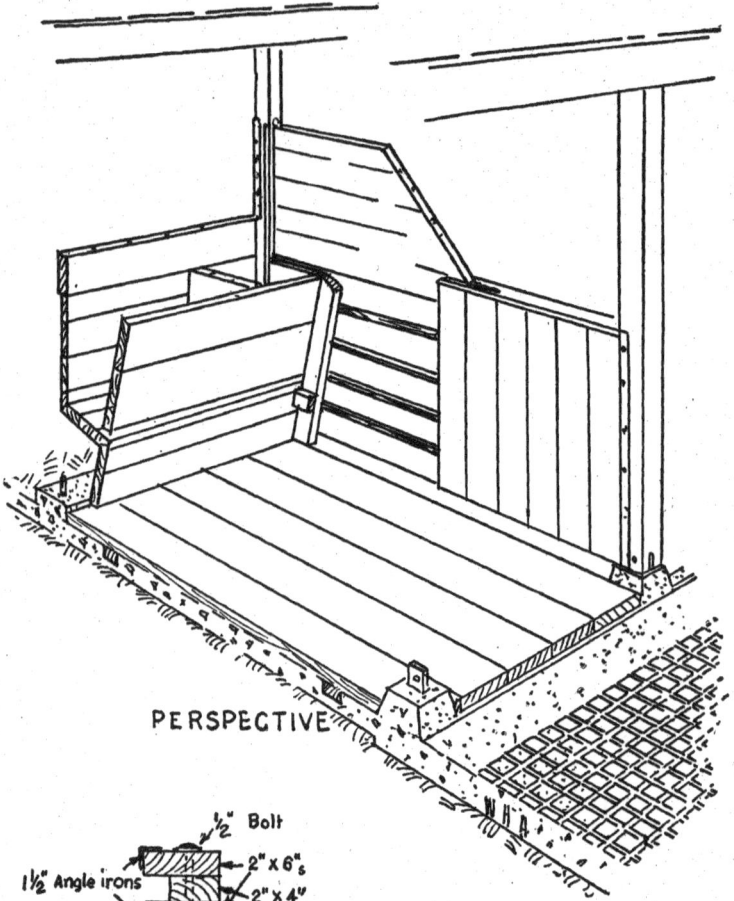

PERSPECTIVE

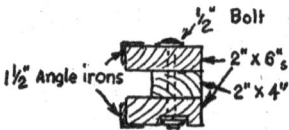

½" Bolt

1½" Angle irons

2" x 6"s

2" x 4"

DETAIL 'A'
Optional construction of posts

(See opposite page for construction plans)

proofed. Minimum size of stall will depend on the animal housed, but the average should not be less than 10 by 12 feet. Ceiling height over horse stall should approximate 8 feet. Avoid the location of ceiling beams directly over the manger tie.

As with all farm-building partition construction, it is well to have the horse stall partitions and the manger frame set on a concrete curb which raises the wood construction at least 6 inches above the floor level. This protects the wood against rot, makes sanitation easier because of the rounded corners, and gives an all-around better appearance. The wood is usually fastened to the concrete curb by means of bolts or splines set in the concrete when poured.

Although work-horse harnesses are often hung on pegs on the stable wall, a better practice is their storage in a separate harness room which can be kept clean, and free from excessive humidity. Since wall space is desired, a long narrow room is preferable to a square one. A tap room in connection with a riding or driving stable should be airy and light, with adequate space for saddles, bridles, and harness and with facilities for washing, cleaning, and oiling all equipment.

Storage Buildings

Silos

Silos are storage structures for holding chopped corn or chopped forage crops cut in a green stage for the feeding of animals, particularly dairy cows. Trench silos are rectangular excavations in a well-drained site, having a sloping end exit, into which the silage is packed. Coverage with straw topped with earth or tar paper and boards gives protection during the use period. Vertical silos are the general rule. These are cylindrical structures of wood, concrete or tile blocks or staves, brick, stone, metal, or of monolithic concrete construction. Any of these materials is satisfactory so long as construction insures walls that are relatively airtight and strong enough to resist the internal pressure of the ensiled material. The diameter of the silo bears a relationship to the size of the herd, and the height to the necessary volume of feed to carry over the silage-feeding months.

The use of green forage crops for silage has emphasized the necessity of adequate reinforcing for silos. Regardless of whether harvesting of grass silage is anticipated, it is well to insist on a reinforcing schedule which meets the requirements of both corn and grass silage. Tests have shown that corn silage under certain conditions exerts pressures comparable to that of grass silage, therefore, no distinction should really be made.

Although additional storage in silos can be secured by extending them underground, the difficulty of removing silage and

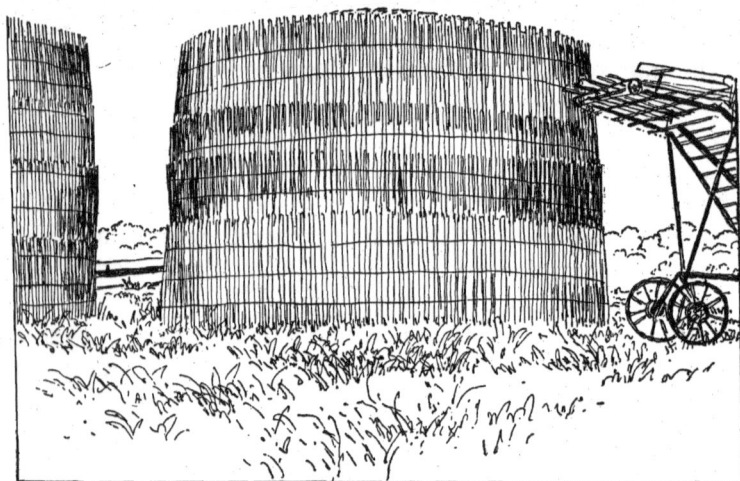

Snow Fence Silo—Temporary Type

trouble with drainage from materials of high moisture content favors placing the silo floor on the feed-alley level. Since the concentrated weight on silo foundation and floor is tremendous, foundation walls must extend to a firm footing and be strong enough to support all possible loads. Wood silos must be firmly anchored to prevent distortion when empty and may require seasonal hoop tension adjustment to compensate for swelling and shrinking of the staves. Painting of wood silos is largely a matter of appearance rather than preservation and, since moisture is moving out through the staves, a paint that is porous will have less tendency to peel and blister due to moisture pressures than one that is impervious. Because of the corrosive action of high-moisture-content silage juices penetrating the silo wall, preference is for placement of reinforcing outside of the wall rather than integral with it. Mortar joints and concrete or plaster surfaces will better resist deterioration if treated periodically with

*Relation of size of herd to diameter of silo for winter feeding, on the basis of 40 pounds of silage per cubic foot and the removal of 2 inches of silage daily to avoid spoilage.**

Inside diameter of silo (feet)	Volume per foot of depth	Amount to be removed—		For a feeding period of—		Animals that may be fed with a daily allowance per head of—			
		Daily		180 days	240 days	40 pounds	30 pounds	20 pounds	15 pounds
	Cubic feet	Pounds		Tons	Tons	Number	Number	Number	Number
10	78.5	524		47	63	13	17	26	35
11	95.0	634		57	76	16	21	31	42
12	113.1	754		68	90	19	25	37	50
13	132.7	885		80	106	22	29	44	59
14	153.9	1,027		92	123	25	34	51	68
15	176.7	1,178		106	141	29	39	59	78
16	201.0	1,340		120	161	33	44	67	89
17	227.0	1,513		136	182	38	50	75	101
18	254.5	1,696		153	203	42	56	85	113
20	314.2	2,094		188	251	52	70	104	139

*From *Farmers' Bulletin 1820*, U.S. Dept. of Agriculture.

Capacity of silos with different diameters and depths of silage[1]

Depth of silage (feet)	Capacity with an inside diameter of—										
	10 feet	11 feet	12 feet	13 feet	14 feet	15 feet	16 feet	17 feet	18 feet	19 feet	20 feet
	Tons	Tons	Tons	Tons	Tons	Tons	Tons	Tons	Tons	Tons	Tons
20	27										
22	30	37									
24	34	41	49								
26	38	46	55	65							
28	43	52	61	72	84						
30	47	57	68	80	92	106					
32	51	62	74	87	100	115	121	148			
34	56	67	80	94	109	125	131	161	180		
36		73	86	101	117	135	142	173	194	216	
38			93	109	126	145	153	186	209	233	258
40			100	117	135	155	165	200	224	249	276
42				124	144	165	177	212	237	264	293
44					152	174	188	224	251	279	310
46						184	198	236	265	295	327
48							209	248	279	310	344
50							220	261	293	326	361

[1]Capacities given are for normal corn silage when the silo is filled at the average speed of 20 to 50 tons per day with 1 man in the silo and refilled once after silage has settled. (From *Farmers' Bulletin 1820*, U.S. Dept. of Agriculture.)

linseed oil, acid-resistant varnish enamels, paraffin wax, or the more penetrating type of bituminous or asphaltic coatings. Metal silos will benefit by treatment with anti-rust compounds such as oils, greases, or paints designed for metal use.

Corn Cribs

The traditional corn crib is V-shaped in cross section, 4 feet wide at the bottom, 6 feet at the top, and as long as necessary to provide the desired capacity. Figure approximately 25 bushels per foot of length. The shape and relative dimensions insure protection against all but driving side rains, provide adequate ventilation and good curing conditions. (See pages 308 and 309 for plan.) Cylindrical storage cribs which are common in some sections are made of perforated metal or open-work tile or masonry construction. Such structures usually have a central ventilating flue, and roof hatches for convenience in filling.

Ordinary corn cribs are best set on concrete or masonry piers capped with a metal shield which extends downward as a guard against climbing rodents. Further protection against rodent damage may take the form of half-inch mesh hardware cloth against the inside surface of the vertical siding and across the bottom of the crib under the floor boards. Floor boards are preferably laid lengthwise to facilitate shoveling.

Common Storages

Every farm should be equipped with storage facilities for fruits, vegetables, and root crops for household use. Underground storages are usually preferred because of the temperature-equalizing effect of the soil. These may be located in outdoor excavations, preferably on a hillside to permit a level entrance, or may constitute a walled-off enclosure having insulated walls in the house basement or cellar. If possible, choose a corner location in the cellar and include or provide a cellar window for ventilation. A dirt floor is preferable to concrete, since it acts as a moisture regulator. Floor bins with slatted bottoms hold potatoes and apples, sand-bottomed bins serve for carrots and beets and for bedding of celery, cabbage, or cauliflower. Shelf space can be provided for jams, jellies, and preserves, for cheese, cured meats, the between-market accumulation of eggs, and for sweet potatoes that have previously been cured at a higher temperature. Squash, pumpkins, and onions are best stored out of the cold storage in a room of higher temperature and lower humidity. Ideal storage room temperatures vary according to the produce, but an average of 40 degrees Fahrenheit and between 60 and 75 per cent relative humidity will give excellent results.

Winter and seasonal storage of farm crops for sale is an excellent practice. It obviates the necessity of dumping produce on the market during the harvest glut, often distributes the labor load more uniformly throughout the season, and gives time for better grading and packaging and consequently a preferred price; there is usually a price advantage in marketing over an extended time period. Such commercial fruit and vegetable storages must be designed according to local requirements; generalization is impossible. Special plans for such storages have been developed by various State Agricultural Experiment Stations and the U. S. Department of Agriculture. These should be consulted before construction starts. Common mistakes are: faulty ventilation or its complete lack; disregard of the fact that fruits and vegetables are alive, breathe, give off moisture and gases which must either be removed or, in the case of moisture, recondensed to maintain humidity and prevent desiccation; and mistaken placement of vapor barriers on the outside of such storage walls instead of the inside, thus tending to load the wall insulation with moisture and cause deterioration. Condensation of moisture in storages maintained at desirable humidities is hardly preventable, but wall and ceiling insulation can be so differentiated that moisture is condensed only on the wall surfaces, thus preventing damaging ceiling drip.

Machinery Storage

The old adage that "a good workman is known by his tools" applies equally to agriculture. Successful farmers take care of their equipment, house it against weathering depreciation, keep it oiled and in good repair. Many farm machines are used only a few days a year and so depreciate more by standing still than by actual use. A separate machine-storage structure is therefore a worth-while investment. Such structures range in type from a long, narrow, open-front shed to a large type of wide building with a center driveway. For convenience in storing equipment,

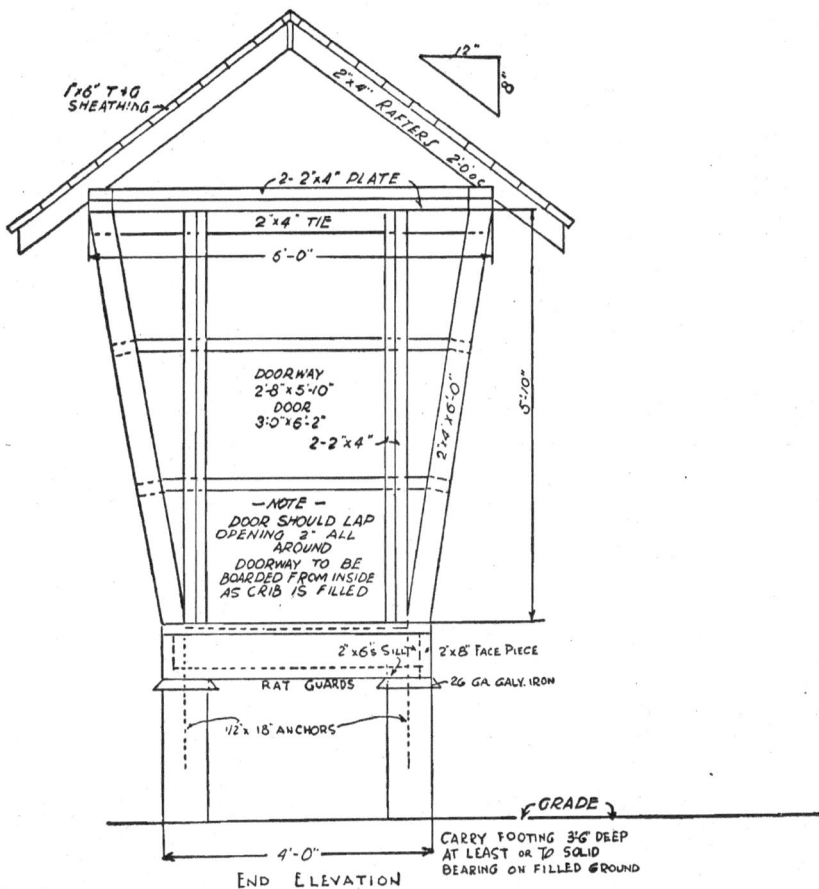

12"
8"

1"x6" T+G
SHEATHING

2"x4" RAFTERS 2'-0" o.c.

2-2"x4" PLATE

2"x4" TIE

6'-0"

DOORWAY
2'-8"x5'-10"
DOOR
3'-0"x6'-2"
2-2"x4"

2"x4"x6'-0"

5'-10"

—NOTE—
DOOR SHOULD LAP
OPENING 2" ALL
AROUND
DOORWAY TO BE
BOARDED FROM INSIDE
AS CRIB IS FILLED

2"x6"s SILLS

2"x8" FACE PIECE

RAT GUARDS

26 GA. GALV. IRON

1/2"x 18" ANCHORS

GRADE

4'-0"

END ELEVATION

CARRY FOOTING 3'-6" DEEP
AT LEAST OR TO SOLID
BEARING ON FILLED GROUND

CORN CRIB
CAPACITY 300 BUSHELS

55

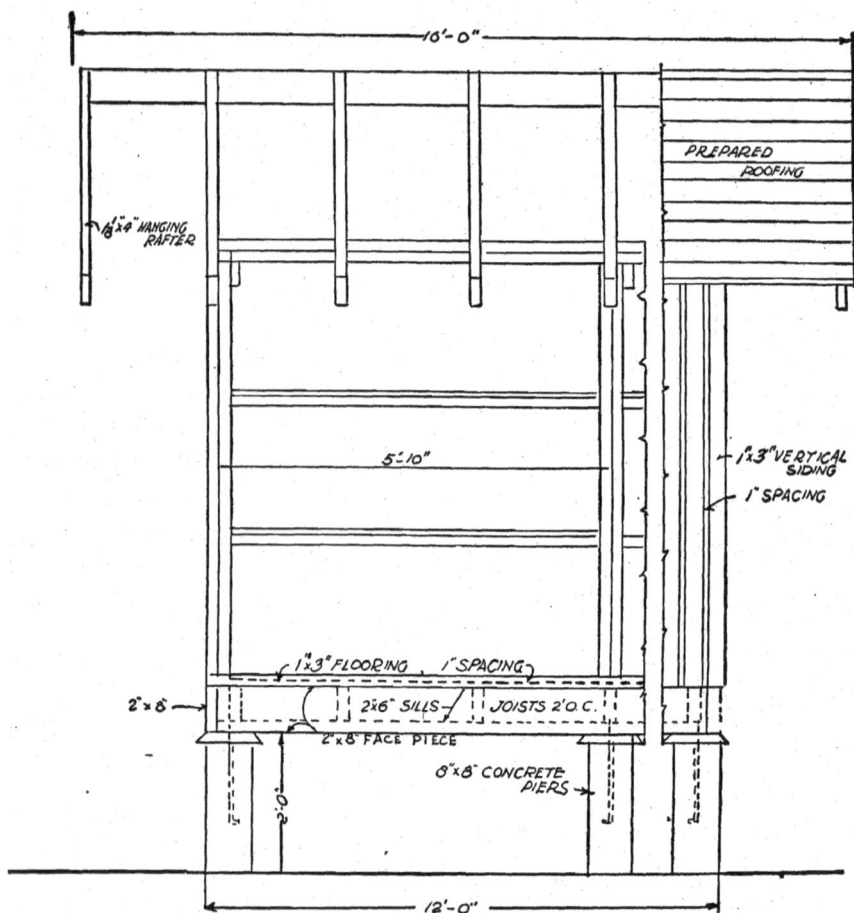

10'-0"

PREPARED
ROOFING

2"x4" HANGING
RAFTER

5'-10"

1"x3" VERTICAL
SIDING

1" SPACING

1"x3" FLOORING   1" SPACING

2"x8"   2"x6" SILLS   JOISTS 2' O.C.

2"x8" FACE PIECE

8"x8" CONCRETE
PIERS

2'-0"

12'-0"

SIDE ELEVATION

Corn Crib may be extended in length as desired to increase capacity.

the former structure is generally preferred. In northern sections particularly, enclosure of the front with a series of sliding doors is recommended. Desirable widths range between 18 and 26 feet; the length varies with the amount of machinery to be stored. It is well to plan generously; few storages are oversize.

Machinery storage construction must be particularly rigid and well anchored to the foundation, because of its shell-like construction and the possibility of wind pressure through open doors. A building of post and nailing girt construction should rest on concrete piers which extend down to solid soil at least below frost line, and above grade from 6 to 12 inches. Side walls and roof of wood construction are common, although corrugated metal and cement asbestos are preferred by many from the fire-hazard standpoint.

Proper repair and reconditioning of equipment is difficult without workshop facilities. Such space is usually provided in one end of the farm machinery storage, separated by a partition and more tightly constructed to facilitate heating during the winter or slack season. A minimum width of 12 feet is suggested with full-width door of height corresponding to the storage entrances. This approximates 10 feet. If possible, the workshop should be equipped with end workbench, a forge, anvil, drill press, and a complement of metal- and wood-working tools in orderly arrangement on wall supports indicated by silhouette outlines. A hoist on movable trunnions or on an overhead track beam will be found very useful. A concrete floor is preferred for the workshop, although an earth or sand-gravel or sand-cinder floor is satisfactory for the storage proper. Windows for the storage are not essential, but good lighting must be provided over the workbench and along the exposed side wall of the shop. Electric wiring for the shop should include a duplex power outlet over the workbench and another in the center of the side wall, a centrally located ceiling light fitted with reflector, and a workbench light of about 100-watt size on a swinging arm or sliding on a ceiling wire parallel to and above the bench.

Garages

Car storage can be provided in one end of the machinery storage if conveniently located to the house driveway. Usually, however, a separate garage structure will be found handier. A very satisfactory one-car garage is of 12′ × 20′ size, with door opening of 8-foot width and 7-foot height. Swinging or sliding doors are least expensive, but the overhead type of spring or weight-balanced door has the advantage that it will not be blocked by snowdrifts or frost-heaved ground. A simple plank workbench and shelves on the rear end wall of the garage will be found very handy. A concrete floor is preferable. A wheel-stop curb 6 or 8 inches high across the rear of the garage prevents damage from careless driving.

A two-car garage should be approximately 20 feet square, with the general construction plan corresponding to that of the one-car structure. The ridge of the gable roof parallels the direction of car entrance and, instead of one window in each of the side walls, two are provided.

Service Buildings

Milk House

Sanitary codes have made the milk house a requirement on dairy farms in most sections of the country. The most convenient location is along the side of the barn, corresponding to the central cross alley. Direct attachment to the barn through a two-doored corridor or passageway is preferred, although some codes require 10 to 20 feet separation between barn and milk house. Consideration should also be given to road access to the milk house for hauling of milk. Furthermore, it should not be located in a barnyard and, if possible, should be surrounded by lawn in preference to the dust and heat of a bare earth yard. Milk houses serve primarily as straining and cooling rooms. These operations require one room. An additional room is desirable if milk utensils are to be washed at the barns instead of in the house.

The following construction is recommended: a minimum width of 10 feet; minimum length of 8 feet, allowing at least 1 foot of length per 24-hour can-production capacity, so as to provide room for the can-cooling tank along the side wall; width of wash room, if added, to correspond with cooling room; minimum length of wash room, 6 feet; level of milk-room floor, 24 to 30 inches above driveway flanking loading platform, to make handling of cans easy; steps or a sloping ramp with a maximum drop of 1 foot in 5, to compensate for the difference in level between barn floor and milk house; cooling and wash rooms should be sheathed and ceiled inside with a non-absorbent, light-colored, and easily cleaned surface, preferably of hard plaster, moisture-proof plywood, pressed-wood, cement asbestos sheets, or metal sheathing; windows should be screened, should represent 10 to 20 per cent of the floor area, and in the wash room should be concentrated over the wash sink location. Doors between cooling and wash rooms should be double-swing, hinge-hung; between milk house and barn, self-closing. Ceiling ventilators extending up through the roof are desirable in both wash and cooling rooms. Floors should be of concrete, with foundation walls extending up above floor level, with a rounded corner fillet to make cleaning easy. A trapped floor drain in each milk-house room is necessary, connected with the discharge lines from both wash sink and cooling tank. Wall construction can be of masonry, poured concrete, or wood framing, with the latter preferred in cold climates because of the possibility of inter-wall insulation fill. Insulation of ceilings is also desirable, particularly to prevent condensation drip.

Where electric water-heating service is available at reasonable cost, this provides the cleanest and least troublesome hot water source for milk-house wash room. Oil, gas, and coal can also be used; in these cases, however, a separate boiler room is often required. With coal, considerable space can be saved and cleanliness assured by locating the sloped bottom bin outside the milk-house wall, with a floor-level shovel entrance conveniently located with respect to boiler.

A frost-proof water supply is necessary and can usually best be run from the barn to the milk house underground through pipe laid inside a 4-inch vitrified bell-end tile line, thus making removal possible without digging up earth or concrete. Copper tubing, if procurable, is recommended for the supply line.

Electric wiring should provide power connection for the refrigerator compressor, duplex convenience outlets at the wash sink for rotary bottle brushes and hot plates, and a reflector-fitted ceiling light in the center of each room and over the wash sink. Specific plans to meet local requirements are usually available through the State Agricultural Extension Service, dairy breed associations, the manufacturers of dairy refrigeration equipment, or building supply representatives.

Feed Rooms

The location and layout of the feed room may make a real contribution in labor saving and convenience. Where feed is processed on the farm, ground or mixed, storage bins for whole grain, for ground feed, and for mixed feed are best located on the second floor, so that advantage may be taken of gravity to supply the grinder, the feed mixer, and eventually the feed carts, thus eliminating the job of hand-handling. Dairy feed rooms are usually located between the silos or adjacent to one, in line with the main feed alley. A slight earth ramp usually makes it possible to reach the second-floor door with a truck load of bagged feed through an outside entrance. A minimum width of 10 feet is suggested, the length being regulated by the feed storage requirements. Bins can be flat-bottomed with chute holes along the wall side; hopper-bottom bins are handier but not worth the extra cost of construction and space wasted.

Feed houses or rooms in connection with poultry plants are usually located at the center of a row of multiple-unit one-story houses or constitute a full-height unit at the end of a multiple-story house. Again, bin storage on the second floor is desirable with gravity feed to mixers and grinders located on the first floor. In the case of multi-story houses, an elevator is a great labor saver. The simplest form is the dumb-waiter type capable of a 500-pound load and hand rope-operated. Where electric service is available, the integral motor-chain hoist is the easiest to install and least expensive. To guide the platform or cage, a 4″ × 4″ vertical rail can be set the length of the hoist reach with the cage or platform center-notched to a loose fit. Self-closing spring-hinged gates are necessary at each landing for accident-prevention.

The storage of large quantities of grain involves structural problems, ventilation, handling, foundation design, and protection against rodents and insects, for which special plans and bulletins detailing various types of construction may be secured through the U. S. Department of Agriculture, Bureau of Agricultural Chemistry and Engineering.

Good equipment is a necessity in American farming practice. Substitution of machines for hand labor characterizes our agriculture, makes for efficient production, reduces drudgery. The well-known slogan of the implement trade that "Good equipment makes a good farmer better" is well founded.

One of the first questions the agricultural neophyte asks refers to the equipment needed for his farm: Typical is the inquiry: "I have just bought a farm of 80 tillable acres; please list the machines I will need—type, make, and price." Such questions cannot be answered without a knowledge of the type of farming intended, crops planned, character of soil. Personal aptitude of the operator may be an important factor in machine choice. The advice of the County or Farm Bureau Agent, the vocational agriculture instructor, the extension agricultural engineer, or of neighbors on whose friendly interest you can depend can be of particular help in the really serious problem of fitting the farm equipment to the farm.

Determination of types and capacity of equipment is of first importance. Following this a choice can be made as to make. The most reliable guide is to choose that equipment handled by a local dealer who is in a position to insure good service with repairs and parts, as evidenced by his reputation. Each particular make of machine may have some desirable features not represented in others, but modern farm equipment is quite uniformly dependable and of high quality regardless of make.

Farm Power Machinery

Field operations are powered either by tractors or draft animals. Those who like to work with animals and appreciate their responsiveness to good treatment and intelligent handling, will find horse or mule power entirely satisfactory and probably a source of enjoyment, especially if the farm operations are limited to one work unit or dependable help can be hired. Belt-power work on such farms is handled with auxiliary internal combustion engines or electric motors.

Those with a mechanical bent who wish to avoid the daily chores of animal care may lean to tractor operation and complete mechanization of the farm. With the advent of the small general-purpose tractor corresponding in capacity to that of a team, mechanization is not limited to the larger farms. A complete range of tractor sizes is available, from the small garden-cultivator type to the heavy-duty unit.

Three distinct types of tractors have been developed to meet the needs of farm conditions. The oldest is the standard four-wheel type, with the centers of the rear drive wheels of about

General-Purpose Tractor Fitted with Cultivator Gangs

48- to 50-inch tread. The front or steering wheel treads usually correspond. Tractors of this type are usually used for plowing, harrowing, for general draft jobs, and for belt work. They are not particularly adapted to row-crop operations, although some are fitted with adjustable rear wheel treads.

The track-laying type of tractor is particularly adapted for work on soft or wet ground and on loose soil. Its ability to turn in a very short radius and its low center of gravity adapt it particularly for rough and hilly land and for land-clearing operations. The general-purpose tractor in recent years has eclipsed all other types in popularity. Usually referred to as row-crop tractors, they are adapted to all types of farming operations. Prevalent types have rear wheel tread adjustable to row-crop spacing, the drive wheels usually straddling two rows, with the single front wheel or dual wheels running in the furrow between. Such tractors are designed for integral mounting of cultivating, harvesting, and seeding equipment—in fact, each manufacturer has developed an entire line of such attaching equipment for, his particular machine. This makes for maneuverability, ease of handling, reduced machinery-investment cost, and places all operations under the immediate control of the driver.

Heavy-duty belt-power machinery is usually operated by the tractor if one is available. In lieu of this, industrial type power units have been adapted for farm use. Many of these are air-cooled to eliminate the bother of a water circulation system, particularly in winter. On horse-operated farms, such engines fill silos, chop hay, grind feed, husk, shred, and shell corn, and saw wood. Where electric service is available, the electric motor represents an ideal power unit. Single-phase service usually permits 5- to 7½-horsepower motors, and these will take care of all but the heaviest jobs. Any hand-crank operation can be handled with a standard quarter-horse motor. It is suggested that these be mounted on a hinged pin or rocking arm, hanging into the belt to keep it at constant tension and facilitating quick detachment for use on another machine. Since adequate wire size is the key to satisfactory electric motor operation, competent electricians should be consulted in determining the layout.

Soil Preparation Machinery

The plow is one of the most universal of farm machines. It turns over, pulverizes, and aerates the soil, and incorporates the vegetation or other organic matter by burying under the furrow slice. Plow sizes refer to the width of furrow slice and range between 6 and 20 inches, the complete designation including the number of bottoms; for example, a 3-bottom 16-inch plow. Although there are hundreds of plow types, they are roughly classified as stubble, general-purpose, and sod-condition plows. The general-purpose is most common and represents a mold-board shape midway between the abrupt pulverizing action of the stubble plow and the smooth gradual turn of the sod mold board. Walking plows have one bottom, may be from 6 to 14 inches in size, and require from 1 to 3 animals. The sulky plow is mounted on a wheel carriage, permitting the operator to ride. Horse-drawn wheel plows having more than one bottom are called gang plows. With a multiple-hitch and a triple-gang plow, one operator can handle a six- or eight-horse team and accomplish as much as a tractor unit in the same number of manhours. Two-way plows are often used for hillside farming. These are essentially two separate plows, one right-hand and one left-hand wheel gang, mounted so that continuous plowing with the furrow turned downhill is possible. A variation of this is the hillside or

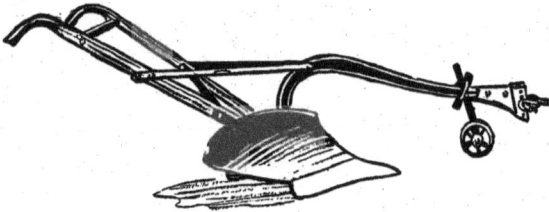

Walking Plow General Purpose

reversible plow, designed with a share having two cutting edges which alternate as share and coulter when the unit is hinged from one side to the other. Due to its shape, this plow does not do a first-class job.

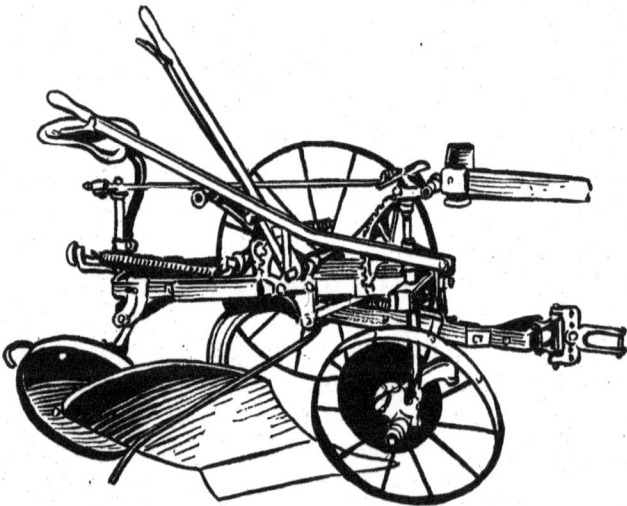

Sulky Plow

Tractor plows attached to the drawbar may be easily wheel-mounted, whether with one or more bottoms. Wheelless tractor plows have been designed for direct attachment to the tractor and incorporate special lift and depth-control devices. These have an advantage in small-tract plowing in that they can readily be backed to the field line, eliminating headlands. All modern

Hillside Plow, Regular Equipment

Tractor Gang Plow

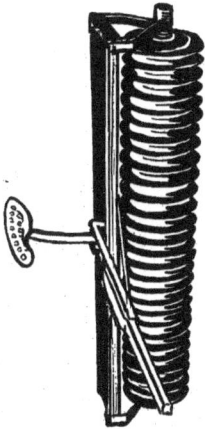

Single-Gang Soil Pulverizer

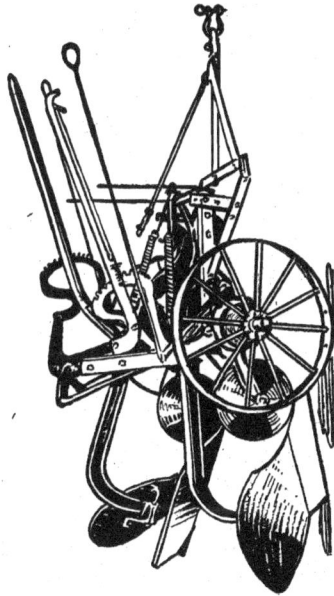

Two-Way Tractor Plow

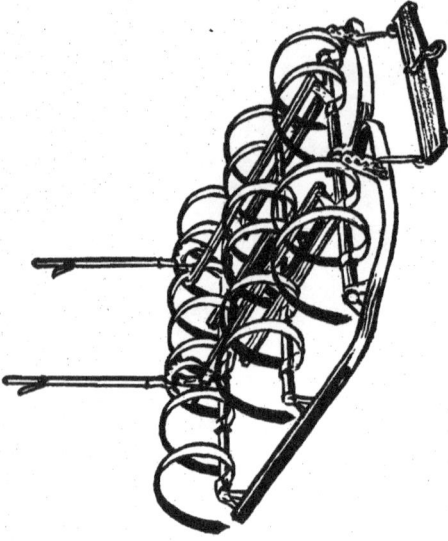

Two-Section 17-Tooth Spring-Tooth Harrow

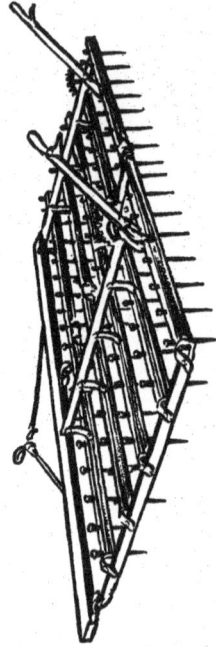

Wood-Frame Peg-Tooth Harrow

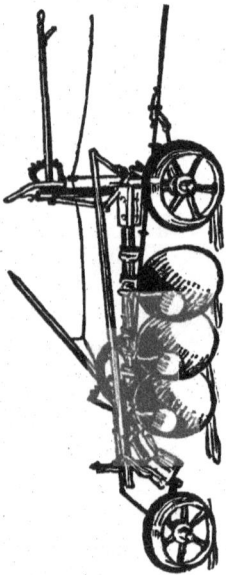

Tractor Disc Plow

Double Disk Harrow (Horse-Drawn)

tractor plows have power lifts and, according to soil conditions, should be fitted with rolling coulter or jointer or both.

A disk plow is essentially a tractor tool designed for extremely hard soils, sticky conditions, or where mold-board plows find penetration difficult.

Harrows are used for smoothing the soil following plowing or for pulverizing the surface of fallow land. Disk harrows are considered most effective for plowed sod land and for pulverizing hard and lumpy soils. Many consider it an essential implement. They are available for both tractor and horse operation, single disk harrows varying between 10 and 20 disks, each covering a width of 5 to 10 feet. Double disk harrows save one operation by having a second gang follow the first one, turning the soil in the opposite direction. Disk blades vary between 14 and 20 inches, with the 16- and 18-inch sizes preferable. The spring tooth-harrow is used in certain sections where firm soil prevents penetration of other types of smoothing harrows. It is particularly adapted to stony soil and for the destruction of deep-rooted weeds such as quack grass. A wide variety of smoothing harrow is offered the farmer. This is a universal implement and finds use on practically every farm in cultivation. It is effective for killing weeds, for smoothing the soil, for planting, for dust mulching, and for leveling. Smoothing harrows consist of a series of pegs mounted on bars which are lever-actuated to adjust the angle of penetration. A special form of harrow used extensively in market garden areas consists of four or more rows of small disks mounted on a frame which slice and compact the seedbed. This is known as Meeker harrow.

Packers and rollers are used to break down lumpy soil and to firm the seedbed. The corrugated roller is preferable to the smooth type, since it leaves the surface slightly roughened and ridged and does a better job of breaking clods. Most effective is the double-corrugated roller having parallel tandem sections. Rollers of this type are considered necessary tools in field-crop soil preparation.

Fertilizer Applicators

Manure spreaders are an essential labor-saving tool on the dairy farm. Recommended practice calls for daily removal of manure from barns to field, whenever conditions permit. A mechanical spreader eliminates the tiresome chore of unloading and spreading by hand, does a much more uniform job, and saves time. Horse-drawn units are four-wheeled; tractor-operated equipment may have four wheels or be fitted with two only and have the front mounted on the tractor drawbar. Horse-

Manure Spreader

drawn units are not necessarily designed to stand up under the rigors of tractor operation. Protection of bearings and working parts from litter and dirt, ease of lubrication and rigidity of construction are points to consider in purchasing. Manure spreaders are rated according to bushel capacity. A 50- to 60-

Lime Sower

bushel spreader is commonly a two-horse unit; tractor spreaders range up to 125-bushel capacity. There is a decided advantage from the draft and depreciation standpoints in having the manure spreader mounted on rubber tires.

Most of our cultivated soils today require applications of lime and commercial fertilizer to supplement manure and organic matter in order to maintain a profitable level of crop production. A great variety of equipment has been developed for special crop and farming needs. Since lime is usually applied in much heavier concentration, equipment for spreading lime is not interchangeable with that for distributing commercial fertilizer. Separate machines are indicated. Lime is applied by means of a broadcast seeder consisting of a box fitted with star-type feed disks supported on two ground wheels. A less expensive method is the use of the end-gate lime sower attached to the rear of a wagon box and chain-driven from a wheel sprocket. This, however, necessitates shoveling from wagon box to sower hopper.

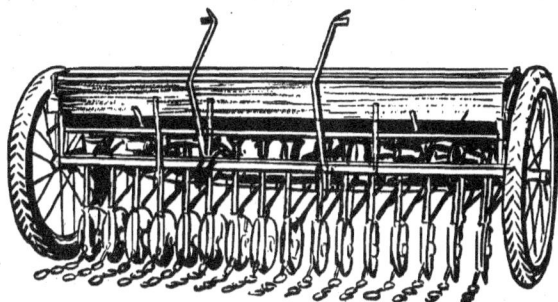

Grain and Fertilizer Drill (with rubber tires)

Fertilizer drills resemble seed drills and have special feed mechanisms which accurately determine the amount of material applied over a wide range of choice, depositing the material broadcast on the ground or in rows made by a furrow opener. For general farming operations, this equipment will be found quite useful, although fertilizer attachments are available and incorporated with standard grain drills and with such equipment as corn planters, potato planters, and various types of cultivating equipment.

Seeding and Planting Equipment

Some type of seeding equipment is desirable even on the small farm. Broadcasting of seeds by hand, particularly grass seed, is a time-honored and fairly satisfactory practice but requires skill born of experience. Much more accurate distribution results from the use of knapsack or wheelbarrow seeders. These are inexpensive and have a wide range of use. Grass seeds need not necessarily be covered following placement, but grain usually requires covering with a mulching harrow or corrugated roller following broadcast seeding. Where a considerable area of field crops is involved, seeders of the standard grain drill type are

FARM STRUCTURES AND EQUIPMENT

designated. These are two-wheeled horse- or tractor-drawn units in varying widths which place the seeds in furrows or can be fitted to broadcast. Standard equipment is available which combines grain seeding, fertilizer distribution, and the surface application of grass or legume seeds. Forage crops are often sown coincidently with grain, so that they can make a first year growth after the grain crop is harvested.

Although cultivated row crops such as corn can be placed with a seed drill, special type planters are usually used for such crops. These can be set to plant in continuous rows or, when used with a check wire, to place the seeds at intervals in hills, so that rows are formed in both directions to facilitate cross cultivation.

Planting of such crops as potatoes and corn can also be done by using a hoe or mattock; hand planters for corn and similar seeds are available and suitable to small areas. Large field operations for any crop require specialized equipment. These needs can best be determined through a study of crop management practices interpreted according to local conditions.

Although most adjustments on drills and planters indicate quantity per acre, variations in seeds make it desirable to test-check equipment before planting. This can be done by raising one drive wheel off the floor, placing a canvas under the seed or fertilizer drops, turning the wheel to correspond to a definite part of an acre traveled in normal field use, and measuring or weighing the quantity of material discharged onto the canvas and comparing it with the discharge indicated by the lever or pointer on the feed adjustment.

Cultivating Equipment

The primary purpose of crop cultivation is the destruction of weeds, though some conditions favor surface cultivation to form a pulverized earth mulch for moisture retention. An infinite variety of tools and equipment has been developed for accomplishing cultivation. For limited crop areas, notably flower and vegetable gardens, hand hoeing is the rule. Handled hoes include the narrow-bladed scraping hoe, a heavier chopping hoe, the multi-tined or finger hoe, and the scuffle hoe which consists of a narrow spring-steel flexible blade supported between prongs. The wheeled hoe is a hand tool which mounts a choice of soil-stirring blades, tines, or shovels on a wheel-supported frame attached to a pair of push handles. The one-wheel jobs are designed for operation between the rows; the two-wheel units straddle the row. Such equipment is frequently used by com-

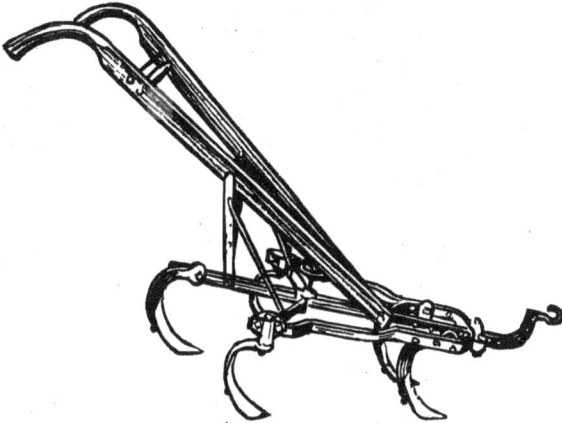

Walking Cultivator

merčial gardeners. Some wheeled hoes combine attachments for seeding and fertilizer application.

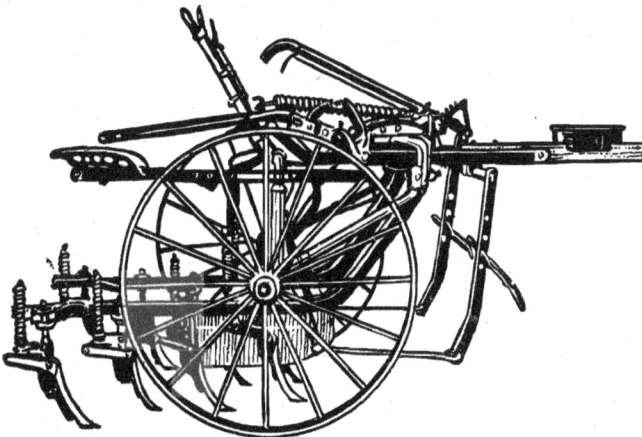

Horse-drawn Straddle Row Cultivator

The one-horse type of walking cultivator is a universal tool for row crops. The soil-stirring elements are usually arranged in the form of a V and the equipment is drawn between the rows, cultivating half of each.

The two-horse riding cultivator or straddle-row type is recognized as the universal row-crop tool for field cultivation. It usually handles but one row at a time, although multi-row units are available.

The general-purpose tractor and the smaller garden tractors have, with cultivating attachments directly mounted on the tractor, largely replaced other cultivating equipment in extensive operations. These usually handle two or more rows, the maximum number being determined by the number of rows seeded or transplanted in one operation. As with the walking and horse-drawn equipment, these cultivator gangs have a variety of shovel,

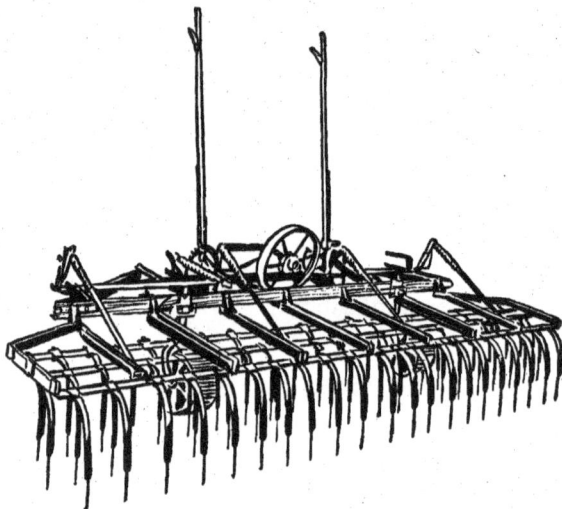

Weeder-Mulcher

sweep, plow, scraper, and tined tools fitted to meet variations in crop, soil, and weed conditions. With the modern high-speed tractor fitted with direct-attached cultivator gangs, one operator can handle 30 to 100 times the acreage possible by hand methods.

The field weeder is a tool rapidly gaining in popularity. This is a wide two-wheeled implement mounting one or more rows of slender spring tines, closely spaced. When used for cultivating growing crops, it is drawn directly over the plants and is very effective in destroying newly sprouted weeds.

Insect and Disease Control Machinery

Disease and insects are an ever-present threat to crop production. The application of insecticides and fungicides in spray or dust form is necessary to the production of high-quality and un-blemished crops. Sprayers can be had in an endless variety of types and sizes. For small garden areas the knapsack unit is effective and inexpensive. Larger areas and the treatment of higher fruit trees may determine the choice of a barrel pump sprayer. This may be mounted on a wheelbarrow or hand truck or transported by wagon. Pumps may be operated by hand or small engine. For field crops and orchards, motor-operated tank-type units are indicated. With them, a better job of pest control can be accomplished because of the higher pressures attainable.

Dusters parallel the range of equipment available in sprayers. For certain crops, insects, or diseases, dusters may be preferable to liquid sprays. For others the spray may be preferable.

Harvesting Equipment

Where hay or other forage crops are raised for harvest, cutting is usually done with a mower. One-horse jobs are available, cutting a swath $3\frac{1}{2}$ to 4 feet wide; two-horse jobs usually cut a 5-foot swath, while tractor-drawn and mounted mowers range to a 7-foot cut. A mower is a necessity where more than an acre or two of material must be harvested.

Hay rakes are used to gather the field-dried forage crops to facilitate loading. The two-wheel dump rake costs least and is most commonly used where hay is hand-loaded, under very hilly conditions and in small irregular fields. The side-delivery rake rolls several swaths together to form a continuous roll, spiral fashion, throughout the field. This facilitates loading with a hay loader, does a cleaner job of picking up, can be utilized for turning damp hay to hasten drying or shake out rain water.

Push rakes are used in some sections to gather hay to a field-located stack or hay storage or to a centrally located baler. These have also proved handy in gathering straw left in the field by a combine.

Hay loaders eliminate the back-breaking chore of lifting forage from ground to transport unit. Several widely varying types are in common use, the most important distinction being between loaders designed for dried material and those for green forage. The latter have been developed for handling green forage crops for ensilage; they are more sturdily constructed and more positive in lift action.

Forage crops are often chopped for easier handling before storage placement. Stationary choppers are tractor- or motor belt-driven, blowing the chopped material directly into the mow where it requires approximately 3/5 of the storage space of loose material. It is more easily removed for feeding and lends itself to better proportioning. Portable field choppers are now available which pick the dried forage directly from the swath or windrow, chop it, and deposit it in adjacent trailer units. At the barn it is blown into the mow by a separate self-feeding blower.

Where forage crops are harvested green for ensiling, all operations have been combined in one machine which mows, elevates, and chops the green material and deposits it in a trailer or separate hauling unit. At the silo it is handled by a separate blower unit.

Where large areas of hay are harvested for farm use and for sale, the field pick-up baler machine has met wide favor. This gathers the hay from swath or windrow and compresses it into bales adjustable between 50 and 100 pounds in weight. Wire bale-tie machines require a minimum of two operators; the twine-tie machine is fully automatic. Both types are operated by power taken from the tractor, or they may be fitted with separate motors.

Grain harvesting equipment is used but a few days a year on the average general-purpose farm and, where investment costs are a factor, such limited harvests are better accomplished by hiring the work done. The grain binder has for years been a universal tool for harvesting seed crops. This cuts the ripe plants, twine-binds them in sheaves, and dumps accumulations of bundles at the will of the operator for setting up into shocks for field curing. Binders for tractor operation are specially designed to withstand more rigorous use; they are driven by power take-off.

The advent of the small combine has made it possible to include the harvesting and threshing operations in a single performance. Such machines cost little more than a power-binder, eliminate the chore of separate threshing, make the farmer independent of his threshing turn, and effect an appreciable saving in labor.

Although special corn binders are available for harvesting green or mature corn, a great deal of this crop is still hand-

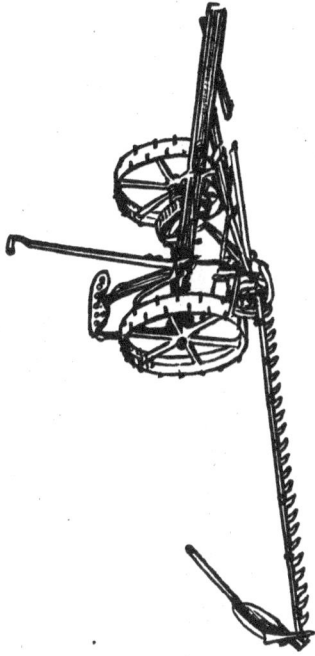

Mower

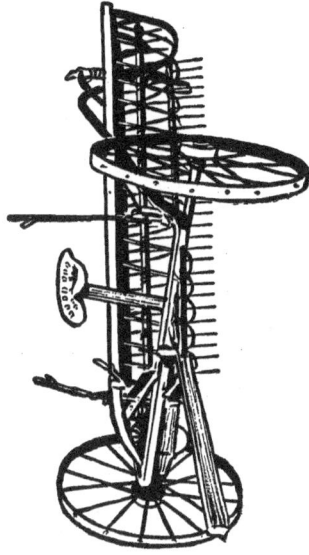

Side Delivery Rake

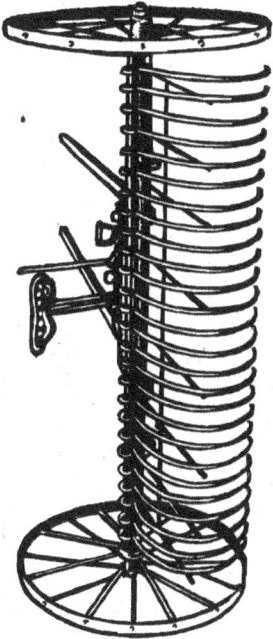

Self-Dump Hay Rake

Grain Binder

Ensilage Cutter

Hay Chopper

Combine

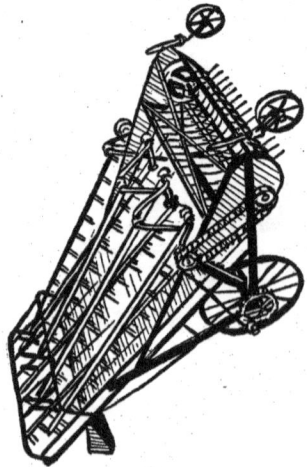

Hay Loader (rake bar type)

harvested with a sickle or corn knife. Small acreages do not justify purchase of special equipment. A silage chopper is, however, necessary in connection with the ensiling of corn. If forage crops are also harvested for silage, the so-called hay chopper can be used for both corn and hay crops. The standard silage cutter does not, however, lend itself well to the chopping of hay.

Special harvesting equipment has been developed to suit particular crops. As a general rule, purchase of such machinery is not justified for small-scale operations. In many cases, the work can be hired to be done by custom outfits for less than the over-

Corn Sheller

all cost of machine ownership. Another satisfactory arrangement is joint ownership with neighbors or the plan whereby two or more neighbors cooperate in each purchasing one machine of several needed by all, thus splitting total machine-investment two or three ways.

Crop Processing Equipment

According to the type of agriculture, each farm will probably require certain processing equipment. Seed graders and cleaners are universally used in grain-growing sections. Although such work can be done by custom rigs or at central mills, grading and cleaning machines cost relatively little and assure the owner a good job of weed-seed removal, separation of impurities, and a grading operation suited to his individual needs. Graders are usually hand-operated and therefore can very readily be fitted with a quarter-horse motor, thus releasing one worker.

Corn shelling is an arduous job if done by hand. Small crank-operated corn shellers are available in one- and two-hole sizes, readily motorized. Where appreciable quantities must be shelled, the cylinder or cannon type of sheller is preferable. This will require tractor engine or heavy motor power.

The grinding of home-grown and purchased grain and subsequent mixing for feeding has many advantages: Quality of the original material is a known factor; usually

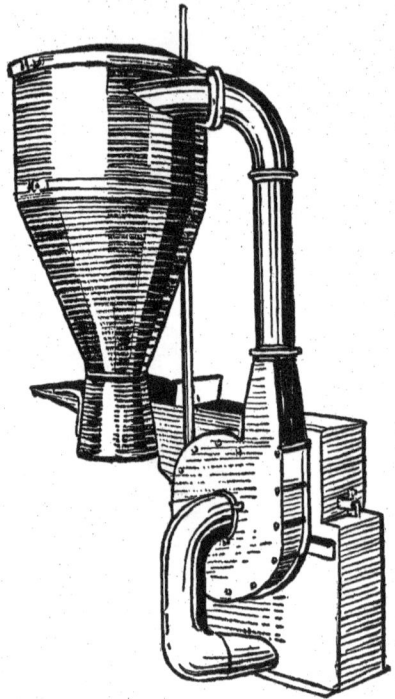

Hammer Mill Feed Grinder with Dust Collector and Bagger

the cost of farm-grown feed is less than that purchased; feed rations can be formulated to fit the use of farm-produced materials with a minimum of purchased supplements. For the average dairy herd or poultry farm, the small automatic-hammer or burr-feed grinder is ideal. Usually electrically operated, the power charge is generally less than the cost of custom grinding; and the bother of bagging and hauling to town and subsequent rebinning operations is eliminated. Since the rate of grinding is necessarily slow, all operations should be automatic: grain should be fed by gravity from overhead bins and ground feed elevated by blower or drag or bucket elevator to overhead bins from which several materials can again be gravity-fed to the feed mixer on the first floor and bagged or discharged directly into the feed cart. From 150 to 1000 pounds per horsepower per hour can be handled with such mills, depending on the fineness of grinding and ‚the kind and condition of grain.

Larger mills for roughage grinding are usually of the hammer type, often including facilities for grain-grinding and mixing. Grinding of roughage, which increases palatability of coarse and low-grade material, is, however, considered unnecessary by many feeding experts.

Dairy Machinery

Many dairymen contend that hand milking still represents the most perfect method. Where help is scarce, however, or unskilled labor is the rule, milking machines are a more satisfactory solution. Properly handled and cared for, such equipment will not injure cows and will produce milk of high quality and low bacteria count; in herds of over 10 to 12 cows, a real saving in labor will be effected. The time necessary to wash and take care of the equipment nullifies other savings in the case of a small herd. Portable machines will handle one or two single or double units; they are plugged into electric outlets conveniently located along the litter alley. Piped-line machines have stationary installed vacuum pumps, the milking pail being separately connected to the pipe system for each cow milked.

Milk Coolers

Milk freshly drawn has a germicidal action which prevents appreciable bacteria multiplication within 3 hours after milking. In many areas, health codes require cooling in less time. Required temperatures are usually 50 degree Fahrenheit or less. Where cold running spring water is available in sufficient quantity, milk in cans immersed in a tank of water will cool satisfactorily. Some dairymen arrange to pump water needed for the stock through the milk cooling tank before it reaches the watering trough, thus accomplishing two purposes. Since water is seldom cool enough, however, ice or mechanical refrigeration is usually necessary to attain the desired low milk-storage temperature. In either case, the water cooling tank should be insulated in the interest of economy. The most satisfactory type of cooling tank is usually made of a sheet metal interior with fill or block insulation surrounding and an outside sheathing of metal, wood, or composition material. Most dairymen prefer the mechanical refrigeration system to ice-cooling, because it eliminates the icing chore, is fully automatic, and insures pre-determined low temperatures. Mechanical cooling units are sized according to the number of 10-gallon cans of milk to be cooled during a 24-hour schedule. Operation costs will approximate one kilowatt hour per can of milk cooled.

Where milk must be cooled for early delivery faster than the 1½ to 2 hours usually required by the immersion method, surface coolers are employed. In these, water or a refrigerant is circulated in a column of horizontal pipe over which the milk is allowed to flow. This results in fast cooling but requires a large reserve

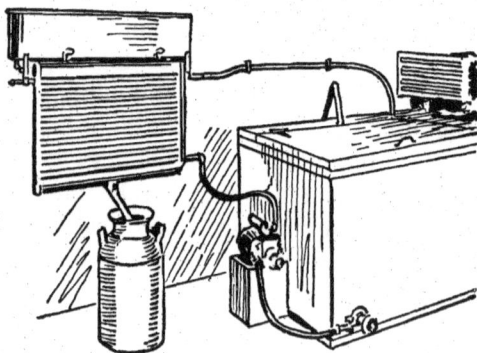

Immersion-type milk-cooling tank with supplementary surface cooler and water-circulating pump.

of cold water brine or refrigerant in order to permit the use of an economically sized compressor unit. There is also the chore of maintaining the surface cooler in a sanitary condition. Although the costs of operation of the two systems of cooling are quite comparable, installation cost of the immersion cooling equipment is considerably less.

Farm Drainage Equipment

The drainage of water from field depressions and from saturated soils in cropped areas is often necessary to avoid the loss of labor, fertilizer, seed, and the time expended in attempted production on such areas. Drainage is equally important in the humid and the arid sections of the country and is especially necessary in some irrigated sections for the removal of excessive ground water accumulations. Drainage is usually by means of agricultural or field tile. These are short tubes, generally 12 inches long, 3 or more inches in diameter, and of various cross-sectional shapes, usually round. They are laid in trenches in lines arranged in herring-bone or parallel pattern and connected together to a common outlet. Such drains lower the water table level below root-damage limits. The depth of the tile trench is determined so as to avoid possible destruction by tillage equipment and heavy machinery, to intercept ground water flows, and to lower the free water level between tile lines sufficiently for crop protection. The tighter and more resistant-to-water flow the soil is, the more closely should the tile lines be spaced. The deeper it is possible to place the lines, the wider will be their effective maximum spacing.

Many factors are involved in the design of a tile drainage system, and those unfamiliar with the practice are advised to consult drainage experts before investing money in trial and error attempts. The cost of drainage varies widely and may range from $25.00 to more than $100.00 an acre. Higher expenditures are warranted in reclaiming potholes, field sumps, and spongy places within cultivated areas, than would be the case for large continuous marshy areas. The nuisance cost of wet spots in otherwise satisfactory fields often exceeds the actual value of the land concerned.

Irrigation Equipment

Natural rainfall is rarely so patterned in frequency and quantity as entirely to meet plant-growth needs. For maximum yields, it is desirable to supplement natural rainfall by irrigation. Where irrigation facilities are costly, the practice must be limited to high-unit-value crops under intensive cultivation.

Regardless of the type of irrigation system used, a regular schedule of watering is essential to the efficient use of equipment and the well-being of the crop. When plants show signs of wilting, it is too late. Crops need water continuously, and systematic watering determined by plant and soil needs in terms of rainfall deficiency is recommended. Ordinary vegetable crops can be satisfactorily handled on a 7-day irrigation cycle. Field crops such as potatoes, corn, and tomatoes require water every 10 to 15 days, while sweet potatoes are often satisfied with a good soaking when vines set and again with a few applications 2 or 3 weeks before harvest. Strawberries and other short-cycle fruiting crops sometimes require irrigation at 3-day intervals during the fruiting period in order to give maximum harvest.

There are several sharply defined irrigation methods. Flood irrigation is used to inundate crop areas suitably leveled. Furrow irrigation places water from a header ditch or pipe line along inter-row furrows of crop and orchard. Overhead sprinkling irrigation utilizes pipe with suitable spray nozzles, covering a strip 25 to 30 feet wide at either side of the elevated pipe line. The line is usually oscillated by means of a water-driven turning motor. This system is particularly favored where land is continuously cultivated in irrigated crops. It is a permanent installation. In a farm management practice where irrigated crops are yearly rotated from one field to another, the portable type of irrigation system is preferred. Several variations are in use. One consists of porous canvas hose from which water seeps to irrigate the rows between which the line is laid. Following a sufficient application, the line is successively moved to adjacent rows. Other methods use lightweight steel tubing with quick-connection couplings, the line fitted at intervals with spray nozzles or gates or, in some cases, with pipe risers terminating in a rotating nozzle capable of irrigating a circular pattern 30 to 60 feet in radius. Such lines are moved by hand progressively across the field.

Operating costs of irrigation systems vary primarily with the source and proximity of water supply. The portable systems are considered somewhat lower in cost of use than the permanent-type overhead sprinkling systems, because of the water volume carried, larger pipes used, and reduced pressure required. Installation costs are commonly less for the portable-type systems than for the permanent, figured on an acreage basis.

Since the manner of pipe layout is of extreme importance in keeping down costs, and because the pumping plant has to be laid out to meet individual conditions, it is recommended that the advice of experienced and competent technicians be solicited. Although initial costs of irrigation equipment may vary widely, thoughtful planning will go a long way in attaining appreciable reductions in maintenance and operating charges.

FUNDAMENTALS OF FARMING

SUGGESTED READINGS

AGRICULTURAL MACHINERY, by J. Brownlee Davidson. (John Wiley & Sons, Inc., New York.)

THE CARE AND MAINTENANCE OF PUMPS AND WATER SYSTEMS. (F. E. Myers & Bro., Ashland, Ohio.)

CONSTRUCTION OF CHIMNEYS AND FIREPLACES, by Wallace Ashby. (*Farmers' Bulletin 1738,* U. S. Department of Agriculture, Washington, D. C.)

DEMING WATER SYSTEMS, SERVICE MANUAL. (Deming Company, Salem, Ohio.)

ELECTRICITY IN THE HOME AND ON THE FARM, by Forrest B. Wright. (John Wiley & Sons, Inc., New York.)

FARM BUILDINGS, by D. G. Carter and W. A. Foster. (John Wiley & Sons, Inc., New York.)

FARM MECHANICS, by F. D. Crawshaw and E. W. Lehmann. (The Manual Arts Press, Peoria, Illinois.)

FARM MOTORS. (Westinghouse Electric & Manufacturing Co., Pittsburgh, Pa.)

FARM PLUMBING, by G. M. Warren. (*Farmers' Bulletin 1426,* U. S. Department of Agriculture, Washington, D. C.)

FARM TRACTORS, by A. A. Stone. (John Wiley & Sons, Inc., New York.)

FARMSTEAD WATER SUPPLY, by G. M. Warren. (*Farmers' Bulletin 1448,* U. S. Department of Agriculture, Washington, D. C.)

HOME ELECTRIFICATION, by R. U. Blasingame. (Edwards Bros., Inc., Ann Arbor, Michigan.)

HOW TO JUDGE A HOUSE. (U. S. Department of Commerce, Washington, D. C.)

MODERNIZING FARM HOUSES, by Wallace Ashby and Walter E. Nash. (*Farmers' Bulletin 1749,* U. S. Department of Agriculture, Washington, D. C.)

POLLUTION OF WELLS AND ITS PREVENTION, by Willem Rudolfs. (*Extension Bulletin 127,* New Jersey Agricultural Experiment Station, New Brunswick, N. J.)

RESERVOIRS FOR FARM HOUSES, by M. R. Lewis. (*Farmers' Bulletin 1703,* U. S. Department of Agriculture, Washington, D. C.)

RURAL ELECTRIFICATION, by J. P. Schaenzer. (Bruce Publishing Company, St. Paul, Minnesota.)

RURAL WATER SUPPLY AND SANITATION, by Forrest B. Wright. (John Wiley & Sons, Inc., New York.)

A SEPTIC TANK DISPOSAL SYSTEM, by E. R. Gross. (*Circular 381,* New Jersey Agricultural Experiment Station, New Brunswick, N. J.)

SEWAGE AND SEWERAGE OF FARM HOMES, by G. M. Warren. (*Farmers' Bulletin 1227,* U. S. Department of Agriculture, Washington, D. C.)

SILOS, TYPES AND CONSTRUCTION, by J. R. McCalmont. (*Farmers' Bulletin 1820,* U. S. Department of Agriculture, Washington, D. C.)

www.ingramcontent.com/pod-product-compliance
Lightning Source LLC
Chambersburg PA
CBHW021604210326
41599CB00010B/602